Disaster Deaths

This book conducts a systematic inquiry into the tragic deaths caused by natural disasters at different geographic scales. It employs key disaster concepts and classification of disasters to understand the high mortality rates and the various factors associated with these deaths.

Deaths are the direct and immediate impact of disaster events, which have remained a major concern for disaster managers and policy-makers all over the world. Using primary research and secondary data, this book provides a comprehensive analysis of various facets of disaster deaths such as trends, circumstances and causes, and determinants at global, regional, national, and subnational scales. It offers a holistic perspective on disaster mortality, which has been lacking for some time. The book not only fills this research gap but also suggests important policy implications for disaster managers and policy-makers working in multilateral, bilateral, local, and international nongovernmental organizations (NGOs). These policies include effective strategies to significantly reduce the risk of deaths caused by natural disasters, which are explored through chapters written in a clear and accessible style. Drawing together the case studies on past major disasters as well as recent ones, the book provides new and critical insights into deaths precipitated by natural disasters.

Suitable for both technical and nontechnical readers, the book has a broader appeal and will thus be useful for practitioners, researchers, students, as well as activists in the area of hazards and disasters who are interested in studying mortality due to extreme natural events.

Bimal Kanti Paul is a Professor of the Department of Geography and Geospatial Sciences at Kansas State University, Manhattan, KS, USA. He is a human geographer with interests in natural hazards and disasters, including human–environment interactions, health and population geography, and geospatial analysis and applications.

Routledge Studies in Hazards, Disaster Risk and Climate Change

Series Editor: Ilan Kelman, *Reader in Risk, Resilience and Global Health at the Institute for Risk and Disaster Reduction (IRDR) and the Institute for Global Health (IGH), University College London (UCL)*

This series provides a forum for original and vibrant research. It offers contributions from each of these communities as well as innovative titles that examine the links between hazards, disasters, and climate change, to bring these schools of thought closer together. This series promotes interdisciplinary scholarly work that is empirically and theoretically informed, with titles reflecting the wealth of research being undertaken in these diverse and exciting fields.

Climate Hazards, Disasters, and Gender Ramifications
Edited by Catarina Kinnvall and Helle Rydström

The Anthropology of Disasters in Latin America
State of the Art
Edited by Virginia García-Acosta

Disaster Resilience in South Asia
Tackling the Odds in the Sub-Continental Fringes
Iftekhar Ahmed, Kim Maund and Thayaparan Gajendran

Crisis and Emergency Management in the Arctic
Navigating Complex Environments
Edited by Natalia Andreassen and Odd Jarl Borch

Disaster Deaths
Trends, Causes and Determinants
Bimal Kanti Paul

For more information about this series, please visit: https://www.routledge.com/Routledge-Studies-in-Hazards-Disaster-Risk-and-Climate-Change/book-series/HDC

Disaster Deaths

Trends, Causes and Determinants

Bimal Kanti Paul

LONDON AND NEW YORK

First published 2021
by Routledge
2 Park Square, Milton Park, Abingdon, Oxon OX14 4RN

and by Routledge
52 Vanderbilt Avenue, New York, NY 10017

Routledge is an imprint of the Taylor & Francis Group,
an *informa* business

British Library Cataloguing-in-Publication Data
A catalogue record for this book is available from the British Library

Library of Congress Cataloging-in-Publication Data
A catalog record has been requested for this book

ISBN: 978-0-367-19626-4 (hbk)
ISBN: 978-0-429-20339-8 (ebk)

Typeset in TimesNRMTPro
by KnowledgeWorks Global Ltd.

Contents

Figures

Tables

Preface

When I began my academic career in Bangladesh in the mid-1970s, my primary focus was agricultural geography, particularly diffusion of agricultural innovations, such as green revolution technology in Bangladesh. After completing advanced degrees in North America in 1988, however, I continued my academic career in the United States as a medical/health geographer. In the late 1990s, my research and teaching interests evolved to include geography of natural hazards and disasters, encompassing topics such as mortality, which overlap with medical/health geography and hazards and disasters geography.

I explained my transition from medical/health geography to geography of natural hazards and disasters in the preface of my first book, *Environmental Hazards and Disasters: Contexts, Perspectives and Management* published in 2011 by Wiley-Blackwell. The devastating floods in 1987 and 1988, the Category 4 tropical Cyclone Gorky that struck the eastern coast of Bangladesh in 1991, and the receipt of several Quick Response Research Grants from the Natural Hazards Center at the University of Colorado at Boulder helped direct my research focus toward natural hazards and disasters. Although I have conducted and published many studies on a wide range of topics associated with disasters, including hazard preparedness, public and household response, individual perceptions and coping strategies to overcome disaster impacts, recovery and rebuilding efforts, and compliance with hazard warnings and evacuation orders, my initial focus was on the disbursement of humanitarian aid to disaster survivors. Based on my writings and practical experience, I authored my fourth book, *Disaster Relief Aid: Changes & Changes* published in 2019.

This book is an amalgamation of my research interests in medical/health, population, and natural hazards and disasters geographies. I have been deeply involved in age- and gender-specific mortality studies, which incorporate perspectives of medical/health geography as well as population and hazards and disasters geography. As a Ph.D. student in the United States, I also was trained in epidemiology, public health, and quantitative techniques. I am also interested in health issues related to extreme natural events, particularly circumstance, causes, and determinants of deaths and injuries caused by floods, tornadoes, tropical cyclones, and earthquakes in Bangladesh, Nepal, the United States, and other countries.

Since 1994, I have received 17 external grants to study natural disasters (e.g., blizzards, cyclones/hurricanes, droughts, earthquakes, floods, and tornadoes), and I have conducted field research throughout the world. Thus, many of my publications are based on primary data. Using secondary data, I have also authored research publications about cyclones, forest fires, heat waves, lightning, tsunamis, and volcanic eruptions in China, Iceland, India, Indonesia, Japan, Sri Lanka, Taiwan, Thailand, and the Philippines. Knowledge gained from my exposure of disaster literature, my involvement in disaster research, and my role as an instructor for an upper level disaster course for nearly two decades was used in this book, including examples of disaster deaths throughout the world. I also taught a disaster course at Jilin University in Changchun, China.

Although this book is not specifically targeted for introductory or upper level interdisciplinary hazard courses, I am confident that undergraduate and graduate students will find it useful in such courses. In addition, emergency managers, planners, government officials, and humanitarian organizations, including nongovernmental organizations (NGOs) and donor agencies involved in disaster death reduction, as well as hazards and disasters researchers, could benefit from this book.

Although the majority of this current book was written during the last one year, my teaching and research experience in three subfields of geography have contributed to the timely completion of the manuscript. Most importantly, its successful completion was contingent on the unconditional support, cooperation, and help I have received from colleagues, friends, and students at Kansas State University (K-State). I am thankful to Dr. Jeffrey Smith, professor of the Department of Geography and Geospatial Sciences at K-State, and Dr. Charles W. Martin, professor and head of the Department of Geography and Geospatial Sciences at K-State, Avantika Ramekar, also at K-State, for drafting all the figures in this book, M. Khaledur Rahman, Assistant Professor in the Geology & Geography Department at Georgia Southern University, Statesboro, Georgia, and Md. Nadiruzzaman, Research Fellow in the Institute of Geography, University Hamburg, Hamburg, Germany. I am particularly thankful to Dr. Martin for allowing me to have access to my office during the COVID-19 pandemic, which allowed me to complete this book project on time. I also received editorial help from Marcella Reekie and Michael Stimers. I would also like to thank Nonita Saha, the editorial assistant at Routledge Geography and Tourism Books for her guidance and patience during this project.

Finally, my deepest gratitude goes to my wife, Anjali Paul, daughters Anjana and Archana, and son Rahul, for their enduring love, constant encouragement, inspiration, and support over the years, particularly when I was busy writing this book.

Bimal Kanti Paul
Manhattan, KS, USA

1 Introduction

Natural disasters are widely conceived as adverse extreme natural events that have detrimental effects on humans and the environment and therefore cannot be considered disasters if they do not harmfully impacts people, or significantly disturb the day-to-day functioning of a community. However, not only can people amplify these harmful impacts, but they are also responsible, to some greater or lesser extent, for creating such damaging events through their activities. For example, floods are generally caused by excessive or prolonged rain over a large area.[1] However, such rain, in turn, may result from human actions, such as deforestation, overgrazing, or drastic land use change (Paul 2011). The disastrous 2017 flooding in Houston, Texas, USA, was caused by a combination of heavy rainfall brought by Hurricane Harvey and unplanned and haphazard development of the city. This Category 4 hurricane not only brought heavy rainfall measured in feet, but also rain-induced water was not able to drain quickly because of sprawling and intense growth of the city. Paving so much ground had reduced its capacity to absorb or rapidly drain rainwater. Flood-control reservoirs in and around Houston not only were too small, but also too few, and building codes were inadequate. As a result, roads became rivers, and neighborhoods became lakes overnight (Coy and Flavelle 2017). In 2005, when Hurricane Katrina made landfall on the Gulf Coast, 80 percent of New Orleans, Louisiana, experienced catastrophic flooding caused by levee failures. The weak levees, which urgently needed repair, could not withstand the force of storm surge water associated with the extremely destructive Hurricane Katrina, which precipitated most of the 1,577 deaths in Louisiana (Levitt and Whitaker 2009; CNN Library 2019).[2]

However, disaster impacts are classified in many different ways. For example, they are popularly dichotomized as having direct and indirect impacts (Smith and Ward 1998). The former or first order impact is the immediate consequence caused by the physical contact of an extreme event with humans and/or with property. Indirect or second order impacts, in contrast, include all those not provoked by the disaster itself, but by its consequences; they manifest much later than the event, and they are often less easily connected to the event (Hallegatte 2015). Both direct and indirect impacts are grouped by tangible and intangible impacts. Tangible impacts can be easily measured in monetary terms, while intangible impacts, which cannot be expressed in monetary terms, include disaster-induced stress, fear, discomfort, and pain (Paul 2011). In order of progression of impacts, Parker et al. (1997) and Smith and Ward (1998) have further trichotomized both direct and indirect impacts as primary, secondary, and tertiary impacts. In essence, primary impacts include immediate impacts, while secondary and tertiary impacts are long-term impacts (Paul 2011).[3]

Another way to express disaster impacts is simply by using three Ds: death, damage, and displacement. The focus of this book is on the first impact and aims to provide a comprehensive analysis of various facets of disaster deaths on global, regional, national, and subnational scales. First, this chapter provides background information to facilitate understanding of the complexities of deaths associated with natural disasters. It starts with a general definition of disaster death, which is often termed as disaster fatality or disaster mortality. Disaster mortality can be defined as deaths that caused by direct and/or indirect exposure of people to an extreme event. Isolating disaster deaths from non-disaster mortality is often a challenging task. Despite several ambiguities associated with disaster deaths, this definition provides a useful and reasonable starting point for analyzing them.

In fact, there are several formal definitions of a disaster death. For example, the United Nations International Strategy for Disaster Reduction (UNISDR) defines disaster deaths as "The number of people who died during the disaster, or directly after, as a direct result of the hazardous event" (UNISDR 2015, 13). In context of flood, Jonkman and Kelman (2005, 75) defined flood fatality or flood-related fatality as "a fatality that would not have occurred without a specific flood event. Synonyms and related terms include 'flood deaths', 'loss of life in floods', 'flood mortality' and 'killed by flooding.'" Applying this to other disasters, disaster deaths are those that would have not occurred without the disaster event.

However, death tolls are widely used to help people conceptualize the magnitude of a disaster. A large number of deaths generally receives the attention of international print and electronic media, which, in turn, is directly tied to the volume of emergency relief to be received by survivors of such an event both from domestic and international sources (Heeger 2007; Paul 2019). Thus, the number of deaths is an important measure of the damage caused by natural disasters.[4]

Deaths and disaster definitions

Death has been included as an important criterion in several definitions of natural hazards and disasters. For example, the Center for Research on Epidemiology (CRED) at the Catholic University of Louvain in Brussels, Belgium, assigns the term disaster if one of the following four criteria are met: 10 or more reported killed; at least 100 or more people reportedly affected (meaning they are in need of immediate assistance from external sources during a period of emergency); call for international assistance; or declaration of a state of emergency (CRED 2015). Several other relevant organizations, such as Desinventar, and Munich Re, the latter a German insurance company, define disaster using the number of deaths caused by such events. Specifically, Munich Re considers a disaster as a situation involving damage and/or loss of lives beyond one million German marks and/or 1,000 persons killed (Dombrowsky 1998).

Similarly, individual researchers have also used mortality in defining natural disasters. Sheehan and Hewitt (1969) defined disasters as events leading to at least 100 deaths, while Glickman et al. (1992) used a lower threshold of only 25 deaths. In the United States, the researchers at Resources For the Future (RFF) have compiled a disaster database with events that occurred between 1945 and 1986, including both natural and industrial disasters. The recording threshold was set at a minimum of 25 fatalities for natural disasters, compared with five for industrial disasters (Smith 2013).

Similar to CRED's EM-DAT, the GLobal IDEntifier Number (GLIDE) maintains an active register of all disasters that have been assigned a GLIDE number, as well as a short description of the event. In order to be assigned a GLIDE number, a disaster must meet the same criteria for EM-DAT (UNEP 2016). Meanwhile, Sigma, a Swiss Re's global disaster loss database, considers an event as a disaster if it causes a certain amount of insured loss in U.S. dollars or results in a certain number of casualties (at least 20 dead or missing, 50 injured, or 2,000 homeless) (UNEP 2016). Unlike EM-DAT, which collects disaster data for hazard researchers and the humanitarian community, Sigma collects disaster data for use by the general public and for the insurance sector in particular. For this reason, Sigma includes a broader range of disasters than EM-DAT (UNEP 2016). Meanwhile, the German insurance industry, Munich Re, and the South American Desinventar consider a disaster as a situation involving damage or loss of life beyond one million German marks or 1,000 persons killed, respectively (Dombrowsky 1998; Aksha et al. 2018).

Deaths and disaster classifications

Apart from defining disasters in terms of number of deaths, often death is used as one of the important components to classify severity of extreme natural events. One of the earliest disaster severity-based classifications was proposed by Jan de Boer (1990). In formulating his classification scheme, he considered seven components or parameters and provided scores for each of the components, ranging from "zero" to "two" (Table 1.1). The sum of the scores lies between 1 and 13 where one reflects very minor disasters and 13 reflects devastating disasters. However, de Boer (1990) did not ascribe any identifier to each composite numerical score or scale, which he calls a Disaster Severity Scale (DSS). Nevertheless, he assigned a scale of 12 for

Table 1.1 Disaster classification by Jan de Boer

Component		
Major	*Minor*	*Assigned value*
Effect (on the surrounding community)	Simple disaster	1
	Compound disaster	2
Cause	Man-made disaster	0
	Natural disaster	1
Duration	Short	0
	Relatively long	1
	Long	2
Radius (of the disaster area)	Small	0
	Relatively large	1
	Large	2
Number of casualties	Minor	0
	Moderate	1
	Major	2
Nature of injuries	No hospitalization	0
	Hospitalization	1
	Large number of severely injuries	2
Rescue time	Short	0
	Relatively long	1
	Long	2

Source: Compiled from de Boer (1990).

the 1988 Armenian earthquake, which killed between 25,000 and 50,000 and injured between 31,000 and 139,000 people. His seven components or parameters and respective scores are presented in Table 1.1.

1. de Boer (1990) considered the *effect on the surrounding community*, with further differentiation into simple and compound effects. In the former case, local and regional rescue services are adequate to deal with the situation. In the latter case, the involvement of national and international organizations is required. A simple disaster is assigned a score of 1, while a compound disaster receives a score of 2 (Table 1.1).
2. de Boer (1990) divides the *cause* into two categories: man-made and natural disasters. He considers that natural disasters are generally more complex and more widely dispersed than man-made disasters. For this reason, the former disasters are accorded a score of one, while the latter get a score of "zero."
3. The next component is the *duration* of an extreme event or how long the event persists in a given area. It is also termed as impact time, and is divided into three categories: short (less than one hour), relatively long (1–24 hours), or long (more than 24 hours). These categories are assigned scores of 0, 1, and 2, respectively.

4 The *radius of the disaster area* is also divided into three categories: small (less than one mile or 1 km), relatively large (1–6.7 miles or 1–10 km), or large (more than 6.7 miles or 10 km). Scores of 0, 1, and 2 are accorded, respectively. This component represents areal extent of physical dimension of disasters, which refers to the area over which a disaster event occurs. This dimension is closely associated with the extent of damage incurred as well as with the number of deaths and injuries (Paul 2011).
5 The fifth component is the *number of casualties* caused by a disaster, which is arbitrarily divided into three categories: minor (25–100 casualties alive or dead, or 10–50 casualties requiring admission to hospitals); moderate (100–500 casualties alive or dead, or 50–250 casualties requiring admission to hospitals); and major (more than 500 casualties alive or dead, or more than 250 casualties requiring admission to "hospitals" (Table 1.1).
6 The nature of the *injuries* sustained by disaster survivors gets a 0 score if no hospitalization is required for injuries. Otherwise, in a typical case a score of 1 is given, unless the disaster results in a relatively large number of seriously injured. In that case, a score of 2 is accorded. Unfortunately, de Boer (1990) did not explain what he means by "large number" (Table 1.1).
7 The *time* required by the search and rescue (SAR) teams to provide first aid to injured people and time to take seriously injured people to appropriate hospital earns a score. This component is also divided into three categories: short (less than 6 hours), relatively long (6–24 hours), or long (more than 24 hours). These categories are graded with a score of 0, 1, and 2, respectively (Table 1.1).

This descriptive severity-based classification of disasters puts emphasis on medical outcomes of disasters, which reflects the bias of the author's medical profession. Of the several physical dimensions of natural hazards (e.g., magnitude, frequency, seasonality, spatial distribution, speed of onset, and diurnal factors), de Boer (1990) only used two dimensions, duration and areal or spatial extent. These physical dimensions most directly affect the number of fatalities caused by a disaster.

Using eight characteristics, one of which is deaths, McEntire (2007) classified extreme events into three classes: crisis, emergency/disaster, and calamity/catastrophe. Instead of using an absolute number, he categorized deaths into three: many, scores, and hundreds/thousands. The remaining seven characteristics are injuries, damage, disruption, geographic impact, availability of resources, number of responders, and time to recover.

Similar to de Boer (1990), Mohamed Gad-el-Hak (2010) classified disasters into five groups. His classification is based on one of two criteria: either number of persons dead/injured/displaced/affected or the size of the affected area of the event. He classifies disaster types as Scope I–V according to the severity scale illustrated in Table 1.2. Although the term "mega-disasters"

Table 1.2 Classification of disaster by Gad-el-Hak

Class (label)	*Number of persons killed, injuries, displaced, or affected*	*Area impacted (in square km)*
Scope I (small disaster)	<10	<1
Scope II (medium disaster)	10–100	1–10
Scope III (large disaster)	100–1,000	10–100
Scope IV (enormous disaster)	1,000–10^4	100–1,000
Scope V (gargantuan disaster)	>10^4	>1,000

Source: Based on information presented in Figure 1 by Gad-el-Hak (2010, 2).

is increasingly used in disaster literature in recent years, particularly since the 1990s when "climate change" started to draw the attention of disaster researchers and others, none of the above classifications used this term. It technically means a high-impact disaster that occurs one-in-a million. Mega-disaster is usually defined as a large-scale disaster that affects five million people, or a one-in-a-million disaster (Sergeant 2011). The CRED (2015) defines a mega-disaster as an event that kills more than 100,000 people. This source identified three mega-disasters in the period 1994–2013. The 2004 Indian Ocean Tsunami (IOT), which killed 226,400 people in 12 countries, is widely considered a mega-disaster. The other two are the 2008 Cyclone Nargis in Myanmar and the 2010 Haiti earthquake. The former killed 138,000 people and the latter 222,600 people (CRED 2015).

Once again, based on de Boer's DSS, Hasani and his colleagues (2014) developed a holistic Disaster Severity Assessment (DSA) tool, which accommodates physical and socio-economic impacts of the disaster on the affected population of a country. It is based on six criteria: (1) impact time, (2) fatalities, (3) casualties, (4) relative financial damage (RFD) in U.S. dollars, (5) Human Development Index (HDI), and (6) Disaster Risk Index (DRI). The last two categories indicate the capacity of the affected country for coping with disaster. The Human development Index has been annually published by the United Nations Development Program (UNDP) since 1990 and is based on three variables: health (life expectancy at birth), education (mean years of schooling), and living standard (gross national income per capita) for each country. The DRI is also available from the World Risk Report published annually by United Nations University. Finally, RFD is calculated by the following formula:

RFD = (original damage of disaster)/(per capita GDP of the affected country)

RFD represents the relative financial damage caused by a specific extreme event in a specific year. The DSA shares three components (time, fatality, and casualty) of the disaster classification of Boer and adds three new parameters (financial damage, HDI and DRI) (Hasani et al. 2014). Similar to de Boer, the sum of the scores ranges from 1 to 13.

Types of disaster deaths

Considering the timing of death caused by natural disasters, these deaths are dichotomized as direct and indirect (Smith and Ward 1998). The former is generally defined as immediate deaths occurring as a result of direct exposure to natural disasters, or deaths directly attributable to the forces associated with disasters. For example, if a person is killed by a building collapse during an earthquake, it can be considered a direct disaster death. Another person may be severely injured during the earthquake, but if the person died a few days or even one year after the event, this can be considered an indirect, secondary, peripheral, or delayed death. Indirect death is caused by the consequences of physical contact of disasters with people (Paul 2011; US DHHS 2017). Thus, disaster injuries provide an important source of information about indirect deaths from such events. Although injury-to-death ratio varies among different types of disasters, and by country to country, the number of injuries experienced across disasters and countries has been approximately twice the number of disaster-related deaths over the last several decades.[5]

Indirect disaster deaths also occur due to outbreaks of infectious/communicable diseases or epidemics in the days, weeks, or even months after the onset of extreme events; however, it is difficult to predict which diseases will occur following an extreme event as they differ by disaster types. For example, water-borne and vector-borne diseases are very common for floods events, particularly in developing countries. Surges of such diseases during and immediately after flood events are a recurrent feature, and people die from these diseases. The widespread prevalence of waterborne diseases is caused by a lack of safe drinking water as floods and other natural disasters often destroy water purification plants, and damage sewage system and sanitation facilities, which, in turn, contaminate drinking and other water sources (Montz et al. 2017). Although flooding flushes away sites for mosquito-breeding and thus reduces incidence of malaria in early stages, subsequently, residual waters may contribute to an explosive rise in the vector reservoir, which, in turn, is associated with increase in incidence of malaria or dengue.[6]

Similarly, problems related to post-disaster rescue and emergency efforts also contribute to an increased number of indirect deaths. The extent of the disaster, the lack of planning for such a disaster, the near total destruction of residential and other buildings, the damage sustained by hospitals, roads, the electrical grid, and other infrastructures, and the fact that medical facilities and professional SAR services and equipment were insufficient even prior to the disaster, all serve to undermine the success of any immediate response. The vast majority of SAR efforts historically have been conducted by friends, family, and neighbors using their hands and commonplace tools. Timely, professional medical treatment has been in drastically short supply; in too many cases, blood loss, infection, renal failure, and other treatable health problems have led to death.

Disaster-induced deaths may also be caused by the transmission of preexisting infectious diseases as well as by preexisting compromised health status of a disaster-affected community, and according to Bourque et al. (2007, 7), "In general, increases in infectious disease rates following disasters are more common in developing than in developed countries." However, lack of access to medical facilities or personnel during and post-disaster periods have generally triggered delayed disaster deaths, and chronic conditions often may be exacerbated by an extreme event. For example, asthma-related deaths are caused by wildfires, and cardiovascular events are associated with blizzards, floods, heat waves, and hurricanes. Both are examples of indirect disaster deaths (US DHHS 2017). In developing countries, disaster-related deaths are exacerbated by lower immunization rates, and poor nutritional status relative to those in developed countries.

Strictly speaking, disaster deaths usually refer to direct deaths. For example, the National Hurricane Center (NHC), located in Miami, Florida, USA, when it provides hurricane fatalities, only reports direct deaths. Thus, actual deaths caused by disasters are higher than the direct deaths, which are generally considered as reported deaths. Irrespective of direct or indirect deaths, disaster-induced deaths can occur before, during, or after any extreme event. People may die even while implementing safety measures before an impending disaster. Moreover, first responders, including police officers and recovery workers also die providing necessary emergency services. These deaths are not considered disaster-related deaths but rather occupation-related deaths (US HHHS 2017). For example, the City of Joplin, Missouri, USA, was hit by a strong EF-5 tornado on May 22, 2011. According to the city, the tornado killed 161 but excluded the death of one police officer who died from lightning while assisting recovery and cleanup efforts the day after the storm (Paul and Stimers 2014).

Thus, direct disaster deaths are deaths that would not have occurred without the occurrence of the disaster event. In contrast, indirect disaster deaths are not caused directly by the forces of the event of the event; such deaths are products of the conditions resulting from the person's contact with the event (McKinney et al. 2011). Accordingly, if a person dies from a heart attack while driving inland to heed a hurricane warning, her/his death is usually considered to be direct. Similarly, when someone dies of a heart attack while cleaning up after a hurricane, that can be considered a direct death caused by the event.

Michael Glantz (2009) argues that disaggregating disaster deaths as direct and indirect is not useful because this diminishes the connection with the event, and thus less may have been done to prevent such deaths (also see Kelman 2005). Further, on his website, Ilan Kelman (www.ilankelman.org/disasterdeaths.html) claims that all disaster deaths, both direct and indirect, are ambiguous. According to him, if a person dies because of drowning in floodwaters, this death can be termed ambiguous. This is because for instance, the basic or fundamental cause of that person's death may be related to poverty which makes the person weak from inadequate food consumption. Or the person dies because of failure of the relevant authorities to issue a warning or because of a lack of rescue. Similarly, a person dies during an earthquake event from a collapsing house, which may not be the inherent cause of the death. Once again, it could be that poverty prevented the person from building his/her house to be resistant to earthquakes. Despite the ambiguity, US DHHS (2017) maintains that for planning and preparedness purposes, recognizing and recording all disaster-related deaths is important, whether the deaths are directly or indirectly related.

Challenges associated with disaster deaths

Sometimes challenges and inconsistencies emerge in the context of disaster deaths or types of disaster deaths. For example, in the case of flood events, perhaps water deaths are confused with flood deaths. In a post, Michael Glantz (2009) provided an example of this confusion.

> "A heart attack during evacuation from a flood is not caused by water; however, it is directly related to the flood disaster. A flood disaster is much more than water. A flood disaster needs water, but it is a primarily disaster event rather than a water event, with all the vulnerability characteristics that a disaster event implies." Similarly, Jonkman and Kelman (2005, 80) claim that deaths can be considered flood-related "only if they occurred during an event involving the presence of water on land that is usually dry. Thus, drownings and heart attacks during floodwater-induced evacuation are flood-related deaths". ....
>
> *(also see Green et al. 2019)*

Further complicating the issue is the fact that many natural disasters arise from a combination of individual geophysical events, or one disaster leading to other disasters (Paul 2011). In other words, one disaster event is connected through a causal sequence to the next extreme event (May 2007; AghaKouchak et al. 2018). For example, earthquakes produce landslides, liquefaction, tsunami, coastal flooding, and fire. In such cases, the disaster that initiated other disasters is called the primary disaster, and the disasters initiated by the primary disaster are most often referred to collectively as secondary or collateral disasters. Both primary and secondary disasters are called cascading disasters. If both primary and secondary disasters cause fatalities, it is often difficult to properly allocate deaths between these two types of disasters. Generally, primary disasters cause more deaths than the secondary disasters. For example, the Boulder, Colorado, USA, area experienced a severe flooding in 2013, which killed eight people. Among eight fatalities, seven were attributed to drowning and one to a flood-induced mudslides (Arnette and Zobel 2016).

Secondary disasters often cause far more damage and problems than does a primary disaster (Montz et al. 2017). For example, the 9.0 magnitude Japan earthquake of 11 March 2011, also known as the Great East Japan earthquake, Tohoku earthquake, or triple disasters, killed more than 18,000 people. Almost all deaths were caused by tsunami generated by the earthquake, which originated off the Pacific coast of Tohoku along a subduction zone 18 miles (29 km) below the sea surface (Pescaroli and Alexander 2015). The massive underwater earthquake triggered tsunami waves that reached heights of up to 133 feet (40.5 m) in Miyako in Tohoku's Iwate Prefecture.[7] These are the high waves that damaged the Fukushima Dai'ichi nuclear reactors whereupon melt down of the reactors led to the evacuation of 200,000 people. The total damage from this cascading disaster was estimated at $300 billion (Karan 2016; AghaKouchak et al. 2018).

Another example of more deaths caused by a secondary disaster as opposed to the primary disaster is the 1960 Chile earthquake, which also triggered deadly tsunami waves. On May 22, 1960, at 3:11 p.m., the country was struck by a 9.5 magnitude earthquake, the world's largest on record. The earthquake originated approximately 100 miles (160 km) off the coast of southern Chile and generated the largest tsunami in the Pacific region for at least the last 500 years. The tsunami was destructive not only along the coast of Chile, but also across the Pacific in Hawaii, Japan, and the Philippines. Tsunami-induced waves up to 82 feet (25 m) high severely battered the Chilean coast. The number of deaths associated with both the earthquake and tsunami in Chile has never been fully resolved; however, estimated fatalities range between 490 and 5,700, with most caused by the tsunami (Pallardy n.d.-b). Thus, the term "secondary" does not refer to scale but rather sequencing (Bradshaw 2013). Yet, for many cascading disasters, it may be difficult to allocate all fatalities appropriately to a single event.

Problems with disaster deaths

The number of disaster deaths has a direct association with media coverage, which not only attracts the world's attention, but also determines the flow of emergency assistance from non-affected countries to the disaster-affected country (e.g., Heeger 2007; Letukas and Barnshaw 2008; Olofsson 2011; Paul 2019).[8] Capitalizing on this, many governments of developing countries often inflate the number of disaster deaths with the intention to garner more foreign emergency aid, or it may be that they are operationally unable to calculate the death count accurately. The reverse may true for developed countries, which do not tend to seek external emergency assistance; often, high mortality is a prestige issue for such countries. As such, they underestimate fatality counts in an effort to hide dire situations.

Also, countries often deliberately underestimate the death toll in an effort to reduce panic among the survivors and others as well as to avoid blame for lack of adequate preparedness measures during the pre-disaster period (Daniell et al. 2013). For example, the 1991 Cyclone Gorky in Bangladesh caused an estimated death of 131,539 persons (Paul 2009). Instead of establishing adequate numbers of public cyclone centers and upgrading the early warning system, both of which could save many lives, the Bangladesh government and its officials blamed the west for the deaths of hundreds of thousands of people. They found an undeniable link between Cyclone Gorky and global climate change. They further claimed that climate change was the consequences of high emissions of greenhouse gases by advanced countries (Chowdhury et al. 1993; Dove and Khan 1995).

Irrespective of level of development, the national government of the disaster-affected countries generally provides an initial estimate of the number of deaths, which needs to revised from time-to-time as more information becomes available.[9] For example, Hurricane Maria, a category 4 hurricane which hit Puerto Rico on September 20, 2017, caused deaths of more than 4,000 people. Puerto Rico is an unincorporated territory of the United States and the government officials initially claimed the disaster caused the deaths of just 16 people. Later, the official estimates revised the number to 64 (Saulnier et al. 2019).

Additionally, a debate raged after the Hurricane Maria struck Puerto Rico. As indicated, as of October 3, 2017, Puerto Rican government authorities reported an official death toll from the hurricane of 64. However, a joint study by the University of Puerto Rico and George Washington University estimated 2,975 excess deaths related to the hurricane in the six months following the event (O'Riley 2018). Their calculation was based on 16,608 carefully reviewed death certificates filed in Puerto Rico between September 2017 and February 2018. Based on household surveying, a Harvard University study put this figure at 4,645, which is nearly 73 times the official total (Arnold 2019). Both studies included both direct and indirect excess deaths in the total number of deaths. Lack of power, pure drinking water and emergency relief aid, inaccessible roads and highways, and mismanagement in handling relief assistance probably caused many indirect deaths after Hurricane Maria made landfall in Puerto Rico. Similarly, elevated deaths statistics have been reported by other researchers for other disasters (e.g., McKinney et al. 2011).[10]

Like most natural disasters, different sources provide different numbers of deaths for the same event; there can be significant disagreement over the exact figure caused by most, if not all, natural disasters. For example, many sources reported the number of deaths for the 2004 IOT, but the death figures differ from one source to another. The number generally increases as the length of reporting increases from the date of occurrence of the event (Paul 2019). For example, on a special issue of the Time on the January 10, 2005, It reported the death toll caused by the 2004 IOT was 123,442 (Time 2005). Almost a year later, Washington Post reported 216,858 deaths caused by the event in its internet site (www.washington.post.com/wp-srv/world/daily/graphics/tsunami_122804.html) on December 22, 2005. At about the same time, the United Nations claimed a death toll of 186,019 (UN 2006).

Similarly, depending on the source, the number of casualties from the 2010 Haiti earthquake differs by an order of magnitude, from a high of 316,000 by the Haitian government, to 222,750 reported by the United Nations to a low of 46,000 determined by an epidemiologic survey.[11] Following the 2010 Haiti earthquake, at least three epidemiologic surveys were conducted and reported estimates from 46,000 to 158,000 deaths.[12] Further a report commissioned by the U.S. government and made public in May 2011 significantly revised the estimate to 85,000. Officials from the U.S. Agency for International Development (USAID) later acknowledged inconsistencies in data acquisition. Given the difficulty of observing documentation procedures in the rush to dispose of the dead, it is considered unlikely that a definitive total would ever be established (Pallardy n.d.-a). This possesses a great challenge with disaster mortality data.

Even in developed countries, counting disaster-induced deaths is a challenging task. Several agencies are responsible for reporting disaster deaths, but their numbers do not always mesh with one another. As death tolls are important in the United States to determine what federal resources are allocated for response, several agencies collect such information immediately after a disaster. Table 1.3, for instance, presents the number of disaster-related deaths for three events occurred in the United States between 2009 and 2012, but the numbers do not correspond with each other. Ideally, death certificates are the primary and most reliable source of official death statistics, but those can take time to appear and do not always include the circumstances of a death. Epidemiological studies, particularly health/mortality surveillance, survey of disaster survivors, funeral homes, coroner's office, or hospital records, are alternative sources of disaster mortality data.[13] Surveillance, which is the systemic collection, analysis, and interpretation of

Table 1.3 Examples of differences in number of disaster-related deaths reported by three different agencies in the United States

	Number of deaths by reporting agency		
Disaster	*Red cross*	*FEMA*[1]	*NOAA–NWS*[2] *storm data*
Hurricane Ike (2009)	38	104	20
Georgia Tornado (2011)	15	9	15
Hurricane Sandy (2012)	34	61	12

[1] Federal Emergency Management Agency
[2] National Oceanic and Atmospheric Administration–National Weather Service
Source: Based on CDC (2017).

deaths, injuries, and illnesses, is a vital source in the process of determining the extent and scope of the health effects of natural disasters on affected populations. However, each source of disaster fatalities information has limitations. Moreover, different methodologies yield different numbers as different agencies have differing guidelines for what to consider in their analysis and calculations. Reliable and complete disaster mortality data are hard to find.

Although the terms excess death or mortality are often used for a particular natural disaster, both are closely associated with indirect disaster deaths. In the context of Hurricane Maria, Carrie Arnold (2019, 23) defines excess mortality as determining the following: "how many people perished in the months following the storm and subtract the number of people who, on average, probably would have died anyway."[14] Although epidemiologists have developed methods to overcome the problem of lack of absolute data, challenges in calculating disaster-induced excess mortality remain. For example, destroyed health-care infrastructure cannot be used for treating disease and illness during the post-disaster period; this leads to excess mortality among disease-stricken people, regardless of whether or not they have been affected by the disaster. It is very difficult to isolate mortality between these two groups of people. Additionally, damaged infrastructure cannot handle the ensuing spikes in disease and illness. But these deaths are most difficult to count (Arnold 2019).

Furthermore, ambiguities exist in defining disaster deaths, even by reputed organizations, such as the CRED, which launched the Emergency Events Database (EM-DAT) in 1988. This database is widely used by hazard and disaster researchers throughout the world because of its reliability and comprehensiveness. The CRED defines disaster deaths as the number of people who lost their life because the event happened. Surprisingly, it includes deaths and persons missing in its count of total deaths caused by a particular extreme natural event. The 2013 World Disasters Report acknowledges that the CRED data on deaths were "missing for around 20 percent of reported disaster" for natural disasters over the last decade (IFRC 2013, 228). The CRED also acknowledges that initially reported data on deaths may sometimes require revision several months later (IFRC 2009).

Often inconsistence exits between identification of a disaster and causes of death. The CRED considered Hurricane Floyd, striking North Carolina, USA, in 1999, and Tropical Storm Allison, which struck Texas, USA, in 2001 as windstorms, even though the majority of fatalities was caused by drowning in inland waterways. Similarly, the 1991 Cyclone Gorky in Bangladesh was classified as a windstorm by EM-DAT, yet most of the deaths were caused by drowning due to storm surges (Haque and Blair 1992; Chowdhury et al. 1993).

Although the absolute number of deaths is widely used to determine the magnitude of a disaster, ideally one should use relative deaths or mortality rates/ratio. The absolute number ignores the scale of the disaster relative to the size of the affected population, which is referred to within the hazards and disasters literature as "impact ratio" (Paul 2011). For example, assume two communities or two spatial units of similar scale (e.g., county or state) experienced 10 disaster deaths in the same event. One community has a total population of 100,000, while other has 10,000 people. In this case, comparing the absolute number of deaths between these two communities is meaningless. More meaningful statistics would be to express total number of deaths per 100,000 or million population and/or deaths per 25,000 square miles (64,750 square km).[14] The former approach is also called the crude mortality rate (CMR). Thus, the number of deaths must be normalized by population or land area, since the impacted area with a larger population, or more land area, may potentially witness more deaths.

Place of disaster deaths

Place or location of death not only differs from one type of disaster to another, but also by national and subnational scales. For example, most earthquake fatalities occur inside buildings, while most flash flood deaths in Europe and North America occur outdoors when people drive through flooded roadways. In contrast, as with earthquakes, most tornado deaths in the United States occur inside buildings of different types: single family homes, mobile homes, apartments, factories, nursing homes, churches, and business buildings. When the location of tornado fatalities in the U.S. is analyzed for 1985–2012, slightly over 41 percent of fatalities occurred in mobile homes, followed by nearly 34 percent in permanent homes. Within buildings, deaths occur in different locations: bathroom, basement, bedroom, garage, hallway, kitchen, living room, and stairwell (Chiu et al. 2013). In the case of 2011 Joplin, Missouri, USA, tornado, other locations for tornado fatalities include about 10 percent in businesses (including hospitals, schools, stores, and churches), slightly over 8 percent in vehicles, and about 5 percent outdoors (Paul and Stimers 2014). Overall, the number and percentage of tornado deaths in mobile homes compared to those in permanent homes is remarkable because mobile homes accounted for only 7.6 percent of U.S. housing units in 2000, and only 6.9 percent of the population lives in mobile homes (Simmons and Sutter 2011).

Place of tornado deaths also varies within a country at different sites in a given year. On April 27, 2011, a record number of 62 tornadoes, including eight EF-4 and three EF-5 tornadoes, struck the state of Alabama, killing 253 people (Chiu et al. 2013). Almost a month later, on May 22, 2011, a deadly EF-5 tornado tore through a densely populated section of Joplin, Missouri, USA, killing 161 people. An analysis of the location of deaths caused by the historic tornado outbreak in Alabama and of the location of all tornadoes that occurred in 2011 in the United States reveals that compared to the Alabama outbreak and nation as a whole, relatively more deaths occurred in business facilities in Joplin (Table 1.4). This is also true when the place of deaths of U.S. tornado victims is considered for the entire

Table 1.4 Number of Tornado fatalities by location, 2011

Location	*USA Tornadoes (excluding Joplin Tornado)* *Number (percentage)*	*Alabama outbreak* *Number (percentage)*	*Joplin Tornado* *Number (percentage)*
Mobile home	112 (28)*	51 (21)	–
Permanent home	164 (41)	148 (60)	65 (41)
Vehicle	19 (5)	11 (5)	15 (10)
Business building	26 (7)	3 (1)	66 (42)
Outside/open	6 (2)	4 (2)	2 (1)
Other/unknown	69 (17)	30 (13)	10 (6)
Total	396 (100)	247 (100)	158 (100)

* Because of rounding, the total percentage may not be 100 percentage.

Sources: Paul and Stimers (2012), SPC (2012), and Chiu et al. (2013).

2000–2011 period. Nearly 35 percent of all deaths that occurred during this period occurred in permanent homes, and only 10 percent occurred in business facilities (SPC 2012). In Joplin, the largest number of deaths occurred in business facilities (i.e., in non-residential structures, such as hospitals, restaurants, churches, and retail stores), followed by permanent homes; this calls into question the protective ability and safety of such structures during tornadoes. More deaths occurred in Joplin in permanent homes and non-residential buildings (e.g., churches, nursing homes, restaurants, hospitals, and retail outlets) relative to the U.S. average primarily because of the lack of a basement in these structures.

Surprisingly, no deaths occurred in Joplin in mobile homes, which typically account for 10–15 times more deaths than do permanent homes in the United States (Paul and Stimers 2014). According to an assessment by the National Oceanic and Atmospheric Administration (NOAA), Joplin residents took shelter after receiving the tornado warning in the most appropriate location (e.g., interior rooms or hallways, or crawl spaces) available to them (NOAA 2011). Even though many residents took action in the final few seconds, the report claims that in many cases, it was a life-saving measure. Unfortunately, below-ground shelters (i.e., basements) are not common in the Joplin area, and some people likely still found themselves in situations that were not survivable (NOAA 2011; NWS 2011).

According to the Jasper County Assessor's Office, nearly 78 percent of houses in the county lack basements, due in part to the rocky ground and high water table (Joplin Globe 2011). Joplin, a city in Jasper county, has an even lower percentage of basements than Jasper County communities as a whole. Also, most of the houses in Joplin are relatively old. In 2009, the median house value in Joplin ($93,108) was 34 percent below the Missouri state average of $139,700 (Joplin, Missouri [MO] Profile 2011). Older houses were constructed according to the standards of the time, which were far less stringent than today's more rigorous building codes. For instance, many of these older houses are not secured to their foundation; some do not even have a foundation (Paul and Stimers 2012).

Not all deaths caused by a particular extreme natural event represent people from the affected areas or communities. For example, of all the deaths caused by the May 22, 2011, Joplin tornado, 138 (85.19 percent) were residents of Joplin; of these 138, 11 (7.97 percent) actually lived outside the damage zones, but at the time of the tornado, they were situated within the damaged zones. Twenty-four tornado victims were residents of 14 neighboring communities of Kansas and Missouri. The relatively high number of non-Joplin-resident deaths reflects Joplin's status as a major regional center lying near the borders of Missouri, Kansas, Oklahoma, and Arkansas. Because the tornado occurred on a Sunday many Joplin residents were away from their homes attending church or high school graduations, visiting friends, shopping, or dining out, among other activities. People came into Joplin that day from neighboring communities for similar reasons, including work (Paul and Stimers 2014).[15]

People died from a particular natural disaster do not always from the affected and non-affected neighboring areas. A deceased person could be from non-affected country for mega-disaster such as the 2004 IOT. The tsunami killed over 200,000 from nearly 50 countries, including non-affected foreign countries. At least 9,000 foreign (mostly European) were among the dead or missing. They came as tourists primarily to Thailand and Sri Lanka. Among the European countries, Sweden suffered the most casualties, whose death was 428, with 116 missing. Other foreign countries experienced a considerable number of deaths were: Argentina, Australia, Austria, Brazil, France, Germany, Japan, Nigeria, Norway, South Africa, South Korea, Switzerland, Ukraine, the U.K., and the United States (Paul 2019).

An analysis of place of disaster deaths is important for three reasons. First, it provides indication of the extent of indirect deaths. For example, if a death occurs in a hospital a few days after the extreme event, it is most likely an indirect death caused by the event. Second, it provides some insight regarding vulnerable structures or space. Lastly, it provides information regarding circumstances of death. In disaster literature, the circumstance of death refers to the timing and place of death (e.g., open space, highways, and residential buildings and non-residential buildings), which differs not only by disaster type but also by a specific type of event occurring over time (SPC 2011). This information is important for implementing mitigation measures as well as for reducing deaths from future such events.

Disaster death studies

The study of disaster deaths is truly multidisciplinary. Many researchers from social, environmental, business, engineering, and health science disciplines are interested in studying disaster deaths temporarily as well as spatially (e.g., Kuni et al. 2002; Kahn 2003, 2005; Jonkman and Kelman 2005;

Noji 2005; Donner 2007; Pradhan et al. 2007; Borden and Cutter 2008; Shapira et al. 2015; Paul and Mahmood 2016). Within a particular discipline, such research is relevant to scholars from more than one field. For example, mortality studies in general, and disaster deaths in particular, are of interest to social scientists of different backgrounds, such as anthropologists, climate scientists, demographers, economists, geographers, geologists, sociologists, and statisticians. Similarly, in the health science discipline, epidemiologists, health care professionals, including trained physicians, and public health specialists, also analyze deaths caused by natural disasters.

Even within a particular discipline, many researchers of different subfields conduct research on disaster deaths. In geography, such analysis of mortality has a direct link with at least four of its subfields: historical geography, hazards and disasters geography, medical/health geography, and population geography. The topic represents one of the three traditional concerns of population geography. Glenn Trewartha (1953), who is considered the pioneer of the sub-field, proposed "a system for population geography." This "system" lists three broad topics to be covered in population geography: historical population geography, population numbers, and qualities of population and their regional numbers. He appropriately identified mortality studies under population numbers.

Mortality also represents one of the two traditional core areas of research in medical/health geography. The first area encompasses disease ecology, which, in the most general sense is the study of interaction between man and his total environment.[16] The principal aim of disease ecology is to understand the dynamics of disease and deaths, which vary according to climate, vegetation, and environment (Paul 1985). Following this tradition, countless mortality atlases have been prepared to examine spatial patterns of cancer, cholera, and other diseases (Borden and Cutter 2008). Despite its direct links with population and medical/health geography, most of the recent work on mortality within geography has been undertaken primarily by hazards and disasters geographers. This subfield is relatively nascent within the broader discipline of geography.[17]

Objectives of this book

Clearly, one of the direct and immediate impacts of natural disasters is deaths; this is universal and has remained a major concern of disaster managers and policymakers all over the world (see Borden and Cutter 2008; Paul 2011). Because of its direct and devastating psychological impact on households, communities, and nations, reducing the number of deaths has been the priority of governments of disaster-prone countries, particularly since the declaration of the 1990s (1990–1999) as the International Decade for Natural Disaster Reduction (IDNDR) by the United Nations (UN).

More recently, the Sendai Framework for Disaster Risk Reduction 2015–2030 (SFDRR or Sendai Framework) was adopted by UN member states in

2015 at the Third UN World Conference on Disaster Risk Reduction held in Sendai City, Miyagi Prefecture, Japan. It is closely associated with the other UN Landmark agreements of the Sustainable Development Goals (SDGs – the 2030 agenda) (Green et al. 2019). The Sendai Framework is a 15-year, voluntary, non-binding agreement that emphasizes that the disaster-prone country has the primary role in reducing disaster risk, along with other stakeholders such as local governments and the private sector. It is the successor instrument to the Hyogo Framework for Action (HFA) 2005–2015: Building the Resilience of Nations and Communities to Disasters, which aims to prevent, prepare for, respond to, and recover from natural disasters as quickly as possible. Its principal objective is to "mainstream and integrate DRR within and across all sectors, including health, and at the same time to evaluate health outcomes from DRR implementation" (Aitsi-Selmi et al. 2015). The Sendai Framework outlined seven global targets and four priorities for action. Its first target is to "substantially reduce global disaster mortality by 2030, aiming to lower the average per 100,000 global mortality rate in the decade 2020–2030 compared to the period 2005–2015" (UNISDR 2015, 12). In this context, this book is not only timely, but it will assist in reaching the first target of the Sendai Framework.

However, the Sendai Framework focuses on the reduction of short-term disaster deaths because this type of death is easy to identify, collect, and report than it is for disaster deaths occurring over a long period of time (Saulnier et al. 2019). It is worthwhile to mention that disaster deaths are often classified as short- or long-term depending on when they occur relative to a disaster's onset. For rapid-onset disasters, most deaths caused are direct and occur in the short-term, while deaths caused by slow-onset disasters accumulate over a longer time (Saulnier et al. 2019).

Although many mitigation and preparedness measures have been implemented across the globe to reduce disaster-induced mortality since the IDNDR, disaster deaths have not significantly decreased. A study by Karlsruhe Institute of Technology maintains that the absolute number of deaths by natural disasters has remained essentially constant with only a slight decrease (KIT 2016). Unfortunately, this decrease occurred in many developed countries, while the number of deaths from disasters has been increasing in some developing countries. Similarly, disaster-induced deaths have decreased for several extreme events, while others show an increasing trend. Further, compared to the global death rate due to all causes, the rate of deaths due to natural disasters has remained quite constant (Chan 2015). Yet, a huge number of disaster deaths occur each year: more than 100,000 per year (Paul 2019).

Disaster deaths not only show considerable year-to-year fluctuation but also vary significantly by country, particularly by a country's level of economic development. Analysis of EM-DAT data shows how levels of development impact a disaster's death tolls. On average, more than three times as many people have died per disaster in developing countries (332 deaths) as in developed countries (CRED 2015; also see Kahn 2005). Among the continents, Asia has experienced the highest number of disaster deaths during the recent decades. Furthermore, irrespective of economic development level, disaster mortality significantly differs within areas of a country. For example, for the United States, Borden and Cutter (2008) reported that the regions of the country most prone to deaths from natural disasters are the South and the Intermountain West (also see Cutter 2001). They also observed a distinctive urban/rural component to the country's patterns of disaster mortality. Similarly, Thacker et al. (2008) claimed that cold-related deaths during the 1979–2004 period were highest in the states of Alaska, Montana, New Mexico, and South Dakota, while heat-related deaths were highest in the states of Arizona, Missouri, and Arkansas.

The total number of deaths caused by natural disasters also differs by disaster type. EM-DAT data show that among natural disasters worldwide, earthquakes (including resulting tsunamis) are typically far more deadly than any other type, claiming 748,621 lives from the period 1996–2015 (CRED and UNISDR 2016). This means that, on average, 37,431 deaths occurred per year, accounting for about 56 percent of the disaster deaths during the above 20-year period – more deaths than all other natural disasters combined. In contrast, flooding is the most common type of disaster (in terms of frequency of occurrence), but it causes a relatively small number of fatalities (Saulnier et al. 2019).

Not only does the number of deaths differ by disaster type, but the causes and circumstances of deaths vary as well. For example, nearly 59 percent of earthquake deaths occurred as a result of the collapse of masonry buildings, and 28 percent were due to secondary effects such as tsunami or landslides (CRED and UNISDR 2016). On the other hand, drowning is the leading cause of flood deaths. Jonkman and Kelman (2005) reported that drowning caused two-thirds of deaths from flooding, while just one-third resulted from physical trauma, heart attack, electrocution, carbon monoxide poisoning, or fire (also see Pradhan et al. 2007; Paul and Mahmood 2016). Another example of variation in causes of death by disaster type is drowning, which rarely occurs during earthquakes, but is a significant cause of death during hurricanes and associated storm surges, tsunamis, and floods (Bourque et al. 2007). Physical dimensions or characteristics (e.g., magnitude, duration, frequency, seasonality, rate of onset, and diurnal factors) of each natural disaster also determine the total number of deaths from extreme events.

Disaster deaths also differ by gender, age, socio-economic condition, ethnicity, race, place of residence, immigration and disability status, and other personal, household, and community characteristics (Juran and Trivedi 2015). As noted, Cyclone Gorky made landfall on the eastern coast of Bangladesh on the night of April 29, 1991 (Paul 2009). A study (Chowdhury et al. 1993) conducted immediately after the cyclone reported that 63 percent of the deaths were children under age 10, who represented only 35 percent of the pre-cyclone population. Also, 42 percent more women than men died. However, the lack of a safe water supply and proper sanitation caused a dramatic rise in the incidence of water-borne diseases, and nearly 7,000 people died from diarrhea, dysentery, and respiratory diseases during the post-disaster period (Chowdhury et al. 1993).

Because of significant loss of life from natural disasters, studies are available on the extent, cause, timing, circumstance, and determinants of disaster mortality. Yet, most of these studies consider single year, specific country, or one type of extreme event. A comprehensive analysis of deaths caused by natural disasters is lacking. Moreover, there have been no attempts to review and summarize the multiplicity of disaster-induced deaths. Currently, not a single book or other type of publication is available that treat disaster deaths holistically (i.e., on different scales, either global or nationally) by individual and also by all types of disaster together for single or multiple years. This book also purports to present disaster deaths in an integrative way, for example, using perspectives of public health, social science, and medical science. Such perspectives are urgently needed to help develop a protocol to prevent more disaster deaths.

The specific objectives of this book are essentially threefold: (1) to analyze levels and trends of disaster death patterns, (2) to examine the causes and circumstances of disaster deaths, and (3) to identify the determinants of disaster deaths. All three objectives are analyzed by disaster type and presented with empirical examples at global, regional, national, and subnational levels. Note that the second and third objectives, while they may appear so, are not identical; the former refers to the medical or pathological causes of deaths, the latter concerns antecedent and predisposing causes and causal factors. Throughout the book, case studies will draw from past major disasters as well as many recent ones across the globe.

Sources of information

Materials presented in this book have been drawn from two primary sources. First, a systematic and rigorous review of a large and often complex body of prior and relevant literature was conducted to derive helpful insights into the three key objectives of this book. Published materials were identified in a computerized literature search of open-source electronic databases that included Scopus, Google Scholar, ISI Web of Knowledge, Web of Science, Medline, PubMed, Science Direct, and Sirius. A number of keywords on each broad and specific topics were entered to perform the searches. Databases were supplemented with both printed published and unpublished works that provided additional information to electronic databases. In addition to pseudo-meta-analyses, primary data and information were collected empirically, as the author has been involved in more than one dozen grants in at least three different countries, studying different aspects of natural disasters, such as cyclones/hurricanes, droughts, earthquakes, floods, riverbank erosions, and tornadoes.

The majority of disaster deaths may be decreased by better planning, warning, and preventive measures that can result from this sort of research (Paul 2011). This research approach is also required to fully clarify the complexity of disaster mortality, particularly when social marginalization and rapid environmental degradation and urbanization have been occurring (Paul 2018). Moreover, UN officials maintain that disaster death tolls could rise in the near future if greenhouse gas emissions are not reduced. Among other issues, this book provides useful insights on how to reduce disaster deaths across the globe, and thus it is an important contribution to the body of knowledge on natural hazards and disasters, adding a significant and novel contribution to existing disaster literature. This book also has historical importance, as hazard and disaster researchers further argue that the best way to reduce disaster deaths is to know who and why people died from such extreme events. It will be of interest to government officials, disaster managers, planners, relief agencies, NGOs, donors, and other development practitioners, both undergraduate and graduate students, and teachers and researchers in the multidisciplinary fields of natural disasters.

Chapter sequencing

This chapter has begun with a brief section on selected definitions of natural disasters wherein human deaths are embedded either explicitly or implicitly in such definitions. Subsequent to this discussion, disaster deaths are presented by disaggregating between direct (immediate) and indirect (delayed) deaths as well as deaths caused by primary (e.g., earthquake) vs. secondary/tertiary events (e.g., tsunami). This is followed by a discussion on the problems of disaster data and location/place of disaster deaths. The scope of multidisciplinary disaster death studies is then briefly discussed. The objectives of the book are presented, and finally, the chapter outline is provided. Eventually, Chapter 1 provides background information to facilitate understanding of the complexities of deaths associated with natural disasters. Chapter 2 explores the possible reasons why some disasters caused excess number of deaths, while others caused surprisingly low counts of fatalities. These reasons are primarily associated with physical characteristics of selected disasters, and preparedness and mitigation implemented by public authorities to reduce deaths from natural disasters.

Chapter 3 addresses the first objective of this book, which is to examine the levels and trends of disaster-induced deaths to find out whether such deaths have decreased or increased in recent decades across all disaster types and geographical scales. A quantitative approach is used to analyze the trends of disaster mortality. Besides, this chapter also discusses two additional topics (one disaster myth, and mass-fatalities and their management) related to disaster deaths. In Chapter 4, a rigorous analysis of (medical) causes of deaths caused by different disasters in different geographic scales is presented, along with specific circumstances that led to those deaths.

In light of the presentation, several recommendations are offered, the implementation of which specifically should reduce disaster-induced fatalities.

What factors actually affect disaster deaths? With the help of available empirical studies and theoretical frameworks, Chapter 5 explores the crucial determinants of disaster deaths by reviewing all types of available longitudinal and non-longitudinal studies (cross-sectional, cross-national, and cross-regional) on determinants of deaths caused by each type of disaster. The final chapter provides a summary of the main findings of this book and a discussion concerning how to reduce deaths resulting from future extreme events is presented by major disaster types. Finally, areas of future research are outlined.

Notes

1. Thunderstorms, storm surges from hurricanes, and rapid snowmelt can also trigger floods. In addition, structural failures of dams and altered drainage behavior, such as the creation of concrete channels, are human activities that can produce a flood.
2. Hurricane Katrina also caused 256 deaths in Alabama, Florida, Georgia, and Mississippi (CNN Library 2019).
3. Generally, no distinction is made among disaster impacts, losses, and damages. Strictly, speaking these terms should not be used interchangeability. For example, with the emergence of climate change literature, researchers are increasingly making distinction between damage and loss. The former typically includes destruction of houses, crops, and infrastructure, while loss is the negative effects of natural disasters that cannot be repaired or restored. However, disaster loss overlaps with disaster damage (Paul 2019).
4. The death toll does not always perfectly equate with the size of the disaster, sometimes large disasters cause a small number of deaths, while small disasters can result large death tolls. Despite this, the death toll is an important indicator of disaster magnitude.
5. Number of injuries or injury rates differs within a country on subnational scales.
6. There are several myths related to disaster deaths, which are discussed in Chapter 3.
7. The height of the tsunami waves exceeded the record set by the 1896 Meiji earthquake, which generated the Sanriku tsunami. The Great East Japan earthquake triggered tsunami waves that travelled up to 6 miles (10 km) inland in the Sendai area. The waves also travelled across the Pacific at a speed of 533 miles (800 km) per hour, reaching from Alaska to Chile. The height of the tsunami wave was recorded at 5.1 feet (1.55 m) at Shemya, Alaska; meanwhile, up to 8 feet (2.4 m) tsunami surges were recorded in California and Oregon, while in Chile, the wave height was 6.6 feet (2 m) (Oskin 2017).
8. National and local media play an important role by requesting the public to donate emergency items such as food, bottled water, and clothing for the disaster survivors. Often media provide the addresses of collection center, which result in the distribution of spontaneous and adequate emergency supplies (see Hettiarachchi and Dias 2013). These necessary supplies contribute in reducing indirect deaths from the disaster.
9. Early casualty estimates after natural disasters are typically not very accurate as they are often based on guesswork. At this stage, it is difficult to accurately

estimate the death toll because of damage and destroy of infrastructure (Alexander 1996; Daniell et al. 2013).

10. Another example can be cited with reference to the number of deaths caused by 1991 Cyclone Gorky in Bangladesh. The Bangladesh government reported an estimated death toll of 131,539 people. But based on a detailed epidemiological study, Chowdhury et al. (1993) estimated a death toll of 67,226, about half the government estimate. This difference is comparable to the estimated deaths caused by the 1970 Bhola Cyclone in Bangladesh. While the government's estimate was that 500,000 people died from the cyclone (Haider et al. 1991), using extensive surveys, Sommer and Mosley (1972) calculated that the Bhola Cyclone killed 224,000 coastal residents.
11. The number of deaths was provided by two official sources: The National Risk Disaster and Disaster Management Agency (SNGRD) and the Haiti's Public Works Department (Centre Nationale des Equipments). As of January 11, 2011, the SNGRD reported a median death toll of 223,469, with a range of 222,570 to 230,000 (Daniell et al. 2013).
12. Based on available casualties report, Daniell and his colleagues (2013) claim that the median death toll is less than half of the official figures provided by the Haiti government. Using a logic tree approach they estimated that the 2010 Haiti earthquake more likely killed 136,933 people, with a range of 121,843–167,082 dead.
13. The Centers for Disease Control and Prevention (CDC) in the United States conducts "death scene investigations" after most natural disasters. The CDC developed comprehensive forms and checklists to record disaster deaths for different types of natural disasters for investigators who collect such information during and after a natural disaster.
14. In disaster context, excess mortality is simply defined as mortality above what would be expected based on number of deaths without the disaster in the population of interest. It is thus mortality that is attributable to the disaster (Green et al. 2019).
15. In cases of tornadoes, deaths are often measured by damage zones or length of the tornado path. The former is applied to a particular tornado, while the latter is applied when death tolls are compared for several tornadoes. For example, Paul and Stimers (2014) analyzed the 2011 Joplin, Missouri, USA, tornado deaths by dividing the damaged area into four zones: catastrophic, extensive, limited, and moderate. These zones were identified based on the horizontal distance from the tornado path. Then death rates were calculated for each of these zones in two ways: death rates per 1,000 people and death rate per square mile.
16. The second major research tradition in medical/health geography focuses on the spatial arrangement and utilization of the principal elements of health care delivery systems, and the characteristics of the population involved (Paul 1985).
17. Similar to geography, within the discipline of epidemiology, a separate subdiscipline, called disaster epidemiology, has recently emerged to the study of the short- and long-term adverse health effects of natural disasters and to predict consequences of future disasters (Songer n.d.). According to the CDC it has three objectives: (1) prevent or reduce the number of deaths, illnesses, and injuries caused by disasters; (2) provide timely and accurate health information for decision-makers; and (3) improve prevention and mitigation strategies for future disasters by gaining information for future response preparation (CDC n.d.).

References

AghaKouchak, A., L.S. Huning, F. Chiang, M. Sadegh, F. Vahedifard, O. Mazdiyasni, H. Moftakhari, and I. Mallakpour. 2018. How Do Natural Hazards Cascade to Cause Disasters? *Nature* 561: 458–460.

Aitsi-Selmi, A., S. Egawa, H. Sasaki, C. Wannous, and V. Murray. 2015. The Sendai Framework for Disaster Risk Reduction: Renewing the Global Commitment to People's Resilience, Health, and Well-being. *International Journal of Disaster Risk Reduction* 6: 164–176.

Aksha, S.K., L. Juran, and L.M., Resler. 2018. Spatial and Temporal Analysis of Natural Hazard Mortality in Nepal. *Environmental Hazards* 17 (2): 163–179.

Alexander, D.E. 1996. The Health Effects of Earthquake in the Mid-1990s. *Disasters* 20: 231–247.

Arnette, A.N., and C.W. Zobel. 2016. Investigation of Material Convergence in the September 2013 Colorado Floods. *Natural Hazards Review* 17(2): 05016001.

Arnold, C. 2019. Death, Statistics and a Disaster Zone: The Struggle to Count the Dead after Hurricane Maria: Intense Controversies Surround Studies of How Many People Perish in Conflicts and Disasters, But Researchers are Developing New Ways to Measure Mortality Rates. *Nature* 566: 22–25.

Borden, K.A., and S.L. Cutter. 2008. Spatial Patterns of Natural Hazards Mortality in the United States. *International Journal of Health Geographics* 17 December (www.ij-healthgeographics.com/content/7/1/64 – last accessed May 25, 2018).

Bourque, L.B., J.M. Siegel, M. Kano, and M.M. Wood. 2007. Morbidity and Mortality Associated with Disasters. In *Handbook of Disaster Research*, edited by H. Rodriguez, E.L. Quaranteli, and R.R. Dynes, pp. 97–112. New York, NY: Springer.

Bradshaw, S. 2013. *Gender, Development and Disasters.* Cheltenham, UK: Edward Elgar.

CNN (Cable News Network) Library. 2019. Hurricane Katrina Statistics Fast Facts, 8 August (www.cnn.com/2013/08/23/us/hurricane-katrina-statistics-fast-facts/index.html – Last accessed August 16, 2019).

CDC (Centers for Disease Control and Prevention). 2017. *Death Scene Investigation After Natural Disaster or Other Weather-Related Events Toolkit: First Edition.* Atlanta: CDC.

CDC (Centers for Disease Control and Prevention). n.d. Disaster Epidemiology: Frequently Asked Questions (FAQs) (www.cdc.gov/nceh/hsb/disaster/faqs.htm#:~:text=Disaster%20epidemiology%20is%20the%20use,of%20disaster%20epidemiology%20are%20to – last accessed June 5, 2020).

CRED (Centre for Research on the Epidemiology of Disasters). 2015. *The Human Cost of Natural Disasters2015: A Global Perspective.* Brussels: CRED.

CRED (Center for Research on the Epidemiology of Disasters) and UNISDR (The United Nations Office for Disaster Risk Reduction). 2016. *Poverty & Death: Disaster Mortality 1996–2015.* Brussels: CRED.

Chan, M. 2015. U.N. Study: Natural Disasters Caused 600,000 Deaths Over 20 Years (http://time.com/4124755/natural-disasters-death-united-nations/ – last accessed August 27, 2018).

Chiu, C.H., A.H. Schnall, C.E. Mertzluft, R.S. Noe, A.F. Wolkin, J. Spears, M. Casey-Lockyer, and S.J. Vagi. 2013. Mortality From a Tornado Outbreak, Alabama, April 27, 2011. *American Journal of Public Health* 103(8): e52–e58.

Chowdhury, A.M.R., A.U. Bhuyia, A.Y. Choudhury, and R. Sen. 1993. The Bangladesh Cyclone of 1991: Why So Many People Died. *Disasters* 17(4): 292–304.

Coy, P., and C. Flavelle. August 31, 2017. Harvey Wasn't Just Bad Weather. It Was Bad City Planning. *Bloomberg Businessweek* (www.bloomberg.com/news/features/2017-08-31/a-hard-rain-and-a-hard-lesson-for-houston – last accessed December 21, 2018).

Cutter, S.L. (ed.). 2001. *American Hazardscapes: The Regionalization of Hazards and Disasters.* Washington, DC: Joseph Henry Press.

Daniell, J.E., B. Khazai, and F. Wenzel. 2013. Uncovering the 2010 Haiti Earthquake Death Toll. *Natural Hazards and Earth System Sciences* 1: 1913–1942.

de Boer, J. 1990. Definition and Classification of Disasters: Introduction of a Disaster Severity Scale. *The Journal of Emergency Medicine* 8: 591–595.

Dombrowsky, W.R. 1998. *Again and Again – Is a Disaster We Call a 'Disaster'? In What is a Disaster?*, edited by Quarantelli, E.L., pp. 19–30. New York, NY: Routledge.

Donner, W.R. 2007. The Political Ecology of Disaster: An Analysis of Factors Influencing U.S. Tornado Fatalities and Injuries, 1998–2000. *Demography* 44(3): 669–685.

Dove, M.R., and M.H. Khan. 1995. Competing Constructions of Calamity: The April 1991 Bangladesh Cyclone. *Population and Environment: A Journal of Interdisciplinary Studies* 16(5): 445–471.

Gad-el-Hak, M. 2010. Facets and Scope of Large-Scale Disasters. *Natural Hazards Review* 11(1): 1–6.

Glantz, M. 2009. Direct Disaster Deaths (www.ilankelman.org/aticles2/fe2004disasterdeaths.pdf – last accessed September 21, 2018).

Glickman, T.S., D. Golding, and E.D. Silverman. 1992. *Acts of God and Acts of Man: Recent Trends in Natural Disasters and Major Industrial Accidents.* Washington, DC: Resource for the Future.

Green, H.K., O. Lysaght, D.D. Saulnier, K. Blanchard, A. Humphery, B. Fakhruddin, and V. Murray. 2019. Challenges with Disaster Mortality Data and Measuring Progress Towards the Implementation of the Sendai Framework. *International Journal of Disaster Risk Science* 10: 449–461.

Hallegatte, S. 2015. *The Indirect Cost of Natural Disasters and an Economic Definition of Microeconomic Resilience.* Policy Research Working Paper 7357. Washington, DC: The World Bank.

Haider, R., A. Rahman, and S. Huq. 1991. *Cyclone'91: An Environmental and Perceptional Study.* Dhaka: Bangladesh Center for Advanced Studies.

Haque, C.E., and D. Blair. 1992. Vulnerability to Tropical Cyclone: Evidence from the April 1991 Cyclone in Coastal Bangladesh. *Disasters* 10(3): 217–229.

Hasani, S., R. El-Haddadesh, and E. Aktas. 2014. A Disaster Severity Assessment Decision Support Tool for Reducing the Risk of Failure in Response Operations. *Risk Analysis* 47: 369–380.

Heeger, B. 2007. Natural Disasters and CNN: The Importance of TV News Coverage for Provoking Private Donations for Disaster Relief. Annual Hazards and Disasters Student Paper Competition. Boulder, Colorado: Natural Hazards Center, University of Colorado at Boulder.

Hettiarchchi, S.L., and W.P.S. Dias. 2013. The 2004 Indian Ocean Tsunami: Sri Lankan Experience. In *Natural Disasters and Adaptation to Climate Change*, edited by Boulter, S., J. Palutikof, D.J. Karoly, and D. Guitart, pp. 158–166. New York, NY: Cambridge University Press.

IFRC (International Federation of Red Cross and Red Crescent Societies). 2009. *World Disaster Reports 2009: Focus on Early Warning, Early Action*. Geneva, Switzerland: IFRC.

IFRC (International Federation of Red Cross and Red Crescent Societies). 2013. *2013 World Disasters Report: Focus on Technology and the Future of Human Action*. Geneva, Switzerland: IFRC.

Jonkman, S.N., and I. Kelman. 2005. An Analysis of the Causes and Circumstances of Flood Disasters Deaths. *Disasters* 29(1): 75–97.

Joplin Globe. May 25, 2011. Sunday's Tornado Upgraded to EF-5 (www.joplin-globe.com/local/x108199255/Sundays-tornado-upgraded-to-EF-5 – last accessed January 16, 2012).

Juran, L., and J. Trivedi. 2015. Women, Gender Norms, and Natural Disasters in Bangladesh. *Geographical Review* 105(4): 601–611.

Kahn, M.E. 2003. The Death Toll from Natural Disasters: The Role of Income, Geography, and Institutions. Mimeo. Medford, MA: Tuffs University.

Kahn, M.E. 2005. The Death Toll from Natural Disasters: The Role of Income, *Geography, and Institutions. Review of Economics and Statistics* 87: 271–284.

Karan, P.P. 2016. Introduction: After the Triple Disaster – Landscape of Devastation, Despair, Hope, and Resilience. In *Japan after 3/11: Global Perspectives on the Earthquake, Tsunami, and Fukushima Meltdown*, edited by Karan, P.P., and U. Suganuma, pp. 1–42. Lexington: University Press of Kentucky.

KIT (Karlsruhe Institute of Technology). 2016. Natural Disasters Since 1900 – Over 8 Million Deaths and 7 Trillion US Dollars. (https://www.sciencedaily.com/releases/2016/04/160418092043.htm - last accessed December 21, 2019).

Kelman, I. 2005. Rights, Responsibilities, and Realities: A Societal View of Civil Care and Security. In *Civil Care and Security*, edited by Gerber, R., and J. Salter, pp. 9–36. Armidale, Australia: Kardoorair Press.

Kuni, O., S. Nakamura, R. Abdur, and S. Wakai. 2002. The Impact on Health and Risk Factors of the Diarrhoea Epidemics in the 1998 Bangladesh Floods. *Public Health* 116(2): 68–74.

Letukas, L., and J. Barnshaw. 2008. A World-System Approach to Post-Catastrophe International Relief. *Social Forces* 87(2): 1063–1087.

Levitt, J., and M.C. Whitaker. 2009. *Hurricane Katrina: America's Unnatural Disaster*. Lincoln, NE: University of Nebraska Press.

May, F. 2007. Cascading Disaster Models in Postburn Flash Flood. In *The Fire Environment – Innovations, Management and Policy: Conference Proceedings*, edited by Butler, B.W., and W. Cook, 443–464. Washington, DC: US Department of Agriculture and Forest Service.

McEntire, D.A. 2007. *Disaster Response and Recovery: Strategies and Tactics for Resilience*. Hoboken, NJ: John Wiley & Sons, Inc.

McKinney, N., C. Houser, and K. Meyer-Arendt. 2011. Direct and Indirect Mortality in Florida during the 2004 Hurricane Season. *International Journal of Biometeorology* 55: 533–546.

Montz, B.E., G.A. Tobin, and R.R. Hagelman, III. 2017. *Natural Hazards: Explanation and Integration*. New York, NY: The Guilford Press.

NOAA (National Oceanic and Atmospheric Administration). 2011. NWS Central Region Service Assessment: Joplin, Missouri, Tornado – May 22, 2011. Kansas City: U.S. Department of Commerce.

NWS (National Weather Service). 2011: NWS Central Region Service Assessment: Joplin, Missouri, Tornado. National Weather Service Assessment (www.nws.noaa.gov/om/assessments/pdfs/Joplin_tornado.pdf. – last accessed February 15, 2012).

Noji, E.K. 2005. Disasters: Introduction and State of the Art. *Epidemiologic Reviews* 27(1): 3–8.

Olofsson, A. 2011. The Indian Ocean Tsunami in Swedish Newspapers: Nationalism after Catastrophe. *Disaster Prevention and Management* 20: 557–569.

O'Riley, A. 2018. FEMA Director Says There's Difference between Direct Deaths and Indirect Deaths' in Puerto Rico, following Trump Remarks, 16 September (www.foxnews.com/politics/2018/09/16/fema-director-says-theres-difference-between-direct-deaths-and-indirect-deaths-in-puerto-rico-following-trump-remarks.html – last accessed September 21, 2018).

Oskin, B. 2017. Japan Earthquake & Tsunami of 2011: Facts and Information. Live Science, September 13 (www.livescience.com/39110-japan-2011-earthquake-tsunami-facts.html – last accessed March 11, 2018).

Pallardy, R. n.d.-a. 2010 Haiti Earthquake. Encyclopedia Britannica (www.britannica.com/event/2010-Haiti-earthquake – last accessed August 3, 2019).

Pallardy, R. n.d.-b. Chile earthquake of 1960 (www.britannica.com/event/Chile-earthquake-of-1960 – last accessed August 12, 2019).

Parker, D., N. Islam, and N.W. Chan. 1997. Reducing Vulnerability following Flood Disasters: Issues and Practices. In *Reconstruction after Disasters: Issues and Practices*, edited by Awotona, A., pp. 23–44. Aldershot: Ashgate.

Paul, B.K. 1985. Approaches to Medical Geography: An Historical Perspective. *Social Science and Medicine* 20(4): 399–409.

Paul, B.K. 2009. Why Relatively Fewer People Died? The Case of Bangladesh's Cyclone Sidr. *Natural Hazards* 50(2): 289–304.

Paul, B.K. 2011. *Environmental Hazards and Disasters: Contexts, Perspectives and Management.* Chichester: Wiley-Blackwell.

Paul, B.K. 2019. *Disaster Relief Aid: Changes and Challengers.* Gewerbestrasse, Switzerland: Palgrave Macmillan.

Paul, B.K., and S. Mahmood. 2016. Selected Physical Parameters as Determinants of Flood Fatalities in Bangladesh, 1972–2013. *Natural Hazards* 83 (2016): 1703–1715.

Paul, B.K., and Stimers, M. 2012. Exploring Probable Reasons for Record Fatalities: The Case of 2011 Joplin, Missouri, Tornado. *Natural Hazards* 64(2): 1511–1526.

Paul, B.K., and M. Stimers 2014. Spatial Analyses of the 2011 Joplin Tornado Mortality: Deaths by Interpolated Damage Zones and Location of Victims. *Weather, Climate and Society* 6(2): 161–174.

Pescaroli, G., and Alexander, D. 2016. A Definition of Cascading Disasters and Cascading Effects: Going Beyond the "Toppling" Domains" Metaphor. *Planet@Risk* 2(3): 58–67.

Pradhan E.K., K.P. West, P.H.J. Katz, S.C. LeClerq, S.K. Khatry, and S. Ram. 2007. Risk of Flood-related Mortality in Nepal. *Disasters* 31(1): 57–70.

Saulnier, D.D., H.K. Green, T.D. Waite, R. Ismail, N.B. Mohamed, C. Chhorvann, and V. Murray. 2019. *Disaster Risk Reduction: Why Do We Need Accurate Disaster Mortality Data To Strengthen Policy And Practice?* New York, NY: UN Office for Disaster Risk Reduction.

Sergeant, A.M.A. 2011. Mega-disasters: Is Your IT Battle-Ready? *The Journal of Corporate Accounting & Finance* 22(5): 3–11.

Shapira, S., L. Aharonson-Daniel, I.M. Shohet, C. Peek-Asa, and Y. Bar-Dayan. 2015. Integrating Epidemiological and Engineering Approaches in the Assessment of Human Casualties in Earthquakes. *Natural Hazards* 78: 1447–1462.

Sheehan, L., and K. Hewitt. 1969. *A Pilot Survey of Global National Disasters of the Past Twenty Years*. Boulder: Institute of Behavioral Science, University of Colorado at Boulder, Colorado.

Simmons, K.M., and D. Sutter. 2011. *Economic and Social Impacts of Tornadoes*. Boston: American Meteorological Society.

Smith, K. 2013. *Environmental Hazards: Assessing Risk and Reducing Disaster*. London: Routledge.

Smith, K., and R. Ward. 1998. *Floods: Physical Process and Human Impacts*. New York, NY: John Wiley & Sons Inc.

Sommer, A., and W.H. Moseley. 1972. East Bengal Cyclone of November 1970: Epidemiological Approach to Disaster Assessment. *Lancet* 299(7759): 1030–1036.

Songer, T.J. n.d. Disaster Epidemiology (www.britannica.com/science/disaster-epidemiology – last accessed June 5, 2020).

SPC (Storm Prediction Center). (2012). 2011 Annual U.S. Killer Tornado Statistics (www.spc.noaa.gov/climo/torn/fataltorn.html – last accessed May 15, 2012.

Thacker, M.T.F., R. Lee, R.I. Sabogal, and A. Henderson. 2008. Overview of Deaths Associated with Natural Events, United States, 1979–2004. *Disasters* 32(2): 303–315.

Time. 2005. Anatomy of a Tsunami. *Time* 165(2): 32–45.

Trewartha, G.T. 1953. A Case for Population Geography. *Annals of the Association of American Geographers* 43(2): 71–97.

UN (United Nations). 2006. *Tsunami Recovery: Taking Stock after 12 Months*. New York, NY: UN.

UNEP (United Nations Environment Programme). 2016. Alternative Classification Schemes for Man-Made Hazards in the Context of the Implementation of the Sendai Framework (www.preventionweb.net/drr-framework/download/w3heir8cl?validate=77336865697238633125 – last accessed September 21, 2017).

UNISDR (United Nations International Strategy for Disaster Reduction). 2015. *Sendai Framework for Disaster Risk Reduction 2015–2030*. Geneva, Switzerland: United Nations Office for Disaster Risk Reduction.

US DHHS (US Department of Health and Human Services). 2017. *A Reference Guide for Certification of Deaths in the Event of Natural, Human-Induced, or Chemical/Radiological Disaster*. Report No. 1. Washington, DC: US DHHS.

2 Reasons for unexpected death toll numbers caused by disasters

While studying the extent of deaths caused by extreme natural events, several disaster researchers (e.g., Chowdhury et al. 1993; Paul 2009; Paul and Stimers 2012; Mersereau 2013) observed across the globe that some events caused excessive numbers of human deaths, while others caused surprisingly few. The idea of too many or too few disaster deaths relative to the immensity of a disaster highlights the associated underlying complexities. Based on a careful literature review, 15 disasters have been selected: nine of them experienced far more deaths than expected while the remaining six experienced far fewer. The criterion used to select the 15 events was much higher or lower than the average annual number of deaths in a country by a particular type of disaster. The average death numbers derived from several decades for a disaster type in a given country and were also considered in selecting these 15 events.

The selected events represent five types of natural disasters: eight hurricanes, cyclones, or typhoons, four earthquakes, and one each of a tornado, flood, and tsunami (Table 2.1). This chapter explores the possible reasons first for the surprisingly large numbers of disaster-induced fatalities and second for the unexpectedly low numbers. The events that caused excessive deaths are presented first by type of disaster followed by the specific events responsible for the few deaths. The lessons learned from these selected disasters can be applied to reduce deaths in future events.

Excessive deaths

1920 Haiyuan earthquake, China

China is considered the most earthquake-prone country in the world followed by Indonesia, Iran, Turkey, and Japan. For example, China experienced 161 earthquakes from 1900 to 2016, the highest number of earthquakes of any country. Most struck in the least populous southwest region of the country, where the terrain is highly mountainous. One of the deadliest was the Haiyuan earthquake. Some claim that the Tangshan earthquake was the deadliest one, which hit the northeastern part of Hebei province on

Table 2.1 List of disasters included by too many or too few deaths

Too many death	*Too few deaths*
Earthquake	
1920 Haiyuan earthquake, China	1960 Valdivia earthquake, Chile
2010 Haiti earthquake	2010 Chile earthquake
Tornado	
2007 Joplin, Missouri, USA Tornado	
Flood	
1889 Johnstown flood, Pennsylvania, USA	
Cyclone/Hurricane/Typhoon	
1970 Bhola Cyclone, Bangladesh	2007 Cyclone Sidr, Bangladesh
1991 Cyclone Gorky, Bangladesh	2009 Cyclone Aila, Bangladesh and India
2010 Cyclone Nargis, Myanmar	2013 Cyclone Odisha, India
2013 Typhoon Haiyan, the Philippines	2015 Cyclone Pam, Vanuatu
Tsunami	
2004 Indian Ocean Tsunami	

July 28, 1976, and killed 700,000 people. But the official death toll is reported at 242,769 (Palmer 2012). This figure is lower than the death toll caused by the Haiyuan earthquake, and for this reason it is selected for this chapter.

The Haiyuan earthquake, also known as "The 1920 Gansu Earthquake," struck rural northern China near Inner Mongolia on the evening of December 16, 1920. According to the U.S. Geological Survey (USGS), it registered a 7.8 magnitude on the Richter scale, but Chinese sources claim the earthquake was of 8.5 magnitude. However, its epicenter was in Haiyuan County in Ningxia Province at a depth of 10 miles (15 km), and the earthquake's strong trembling lasted for ten minutes. The earthquake caused total destruction in the Lijunbu-Haiyuan-Ganyanchi area and was assigned the highest destructive intensity of XII on the Modified Mercalli intensity scale (MM or MMI).[1] It rocked the neighboring Gansu and Shaanxi provinces. Furthermore, it was felt as far away as Norway and was followed by a series of aftershocks for three years. Due to the rigidity and fragility of loess, the earthquake triggered 675 major landslides and created ground cracks throughout the epicentral area. The earthquake occurred in the most extensive stretch of loess terrain on Earth (Fuller n.d.).

The Haiyuan earthquake killed 200,000–270,000 people and ranked the fourth deadliest earthquake and ninth deadliest natural disasters in China. About 50 percent of the fatalities were caused by landslides. Over 73,000 deaths were reported in Haiyuan County alone, accounting for 50 percent of the county's total population. The earthquake also killed more than 30,000 in nearby Guyuan County, whilst a landslide buried the town of Sujiahe in Xiji County (Kte'pi 2011). The principal reason for the extremely high death toll was the collapse of cave dwellings. As indicated, the earthquake struck the extensive loess terrain of China where most people live in such dwellings constructed of dug out loess, a type of dwelling highly susceptible to collapse by earthquakes. Many homeless survivors also died because of both hunger and exposure to the windstorms and snowfall that immediately followed the earthquake. In addition to human deaths, over one million cattle, sheep, and other farm animals also died (Fuller n.d.). A large area (19,000 square miles or 50,000 square km) experienced XII destruction on the MMI, which also contributed to a huge number of fatalities.

2010 Haiti earthquake

On January 12, 2010, at 4:53 pm, a tragic 7.0 magnitude earthquake on the Richter scale struck Haiti near Port-au-Prince.[2] The earthquake caused estimated death tolls of 222,750, and injured 300,000 others (Kirsch et al. 2012).[3] The death toll of the Haiti earthquake was much higher than the number of annual average deaths caused by earthquakes globally between 1996 and 2015. Calculations based on data in Emergency Events Database (EM-DAT), developed by the Center for Research on the Epidemiology of Disasters (CRED), reveal that from 1996 to 2015, 37,431 deaths occurred per year (CRED and UNISDR 2016).

Estimates suggest that approximately three million of Haiti's nine million people have been affected by the earthquake. More than 300,000 homes collapsed and more than 700,000 people were displaced (Paul 2019). Fifty percent of buildings were demolished in Port-au-Prince, the capital of Haiti, where more than two million people lived in its metropolitan area. Most of the buildings/houses in the capital city were self-built, without foundations and seismic-resistant features. Many houses were built on steep hillsides, not on solid rock but on lose soil, which collapses when shaken. Unknown numbers of residents of the capital city were crushed inside the buildings. The consequences of this type of building collapse were disastrous and the main reason for the deaths of so many people. Studies conducted on search and rescue operations after the earthquake suggest that about 50 percent of people buried under collapse buildings survived two to six hours after entrapment. Haiti did not have the resources to quickly rescue those people (Daniell et al. 2013). In addition, Haiti had no emergency medical team to save some people who required medical help immediately after the earthquake since the country had financial difficulties. Moreover, Haiti was not at all prepared for the event. When preparedness is lacking, disasters generally kill more people. Some experts claimed that the government was not prepared because the country is perceived to be at low risk for earthquakes. According to the USGS, the last earthquake occurred in the country in 1994. This 5.4 magnitude event killed only four people.

However, the 2010 earthquake was soon followed by two aftershocks of magnitudes 5.9 and 5.5. More aftershocks occurred in the following days, including another one of magnitude 5.9 that struck on January 20 at Petit Goâve, a town some 35 miles (55 km) west of Port-au-Prince. In all, 59 aftershocks were recorded each registering greater than 4.5 on the Richter scale

(Coles et al. 2012). In the aftermath of the event, emergency disaster relief goods, such as food, water, and medical supplies, started to pour in from foreign countries, but were hampered by the failure of the electric power system, loss of communication lines, and roads blocked with debris.

> A week after the event, little aid had reached beyond Port-au-Prince; after another week, supplies were being distributed only sporadically to other urban areas. Operations to rescue those trapped under the wreckage had mostly ceased two weeks into the crisis.
>
> *(Pallardy n.d)*

Because many hospitals and medical clinics had been either destroyed or damaged, seriously injured earthquake survivors had to wait for days, which caused additional deaths. Deaths were also caused by unhygienic living conditions in densely packed temporary shelters. Two years later, more than half a million people remained in tents, many of which had deteriorated significantly (Pallardy n.d.). Ten months after the earthquake, a cholera epidemic surfaced and reached to the tent cities of Port-au-Prince, which caused a considerable number of indirect deaths.

The physical dimensions of earthquakes such as magnitude, intensity, depth of epicenter, and epicentral distance to densely populated areas are associated with the extent of mortality. The magnitude of the Haiti earthquake was not considered "high," but its epicenter was close to the capital city – only 10 miles (15 km) southwest of Port-au-Prince. The earthquake's hypocenter, or focal point, was close to the surface – only at a depth of 8.1 miles (13 km). At such a shallow depth, both primary and shear-waves travel relatively quickly and hence allow less warning time to get out of buildings than do deep earthquakes. If the earthquake had happened further down, it would have lost its energy as it moved up, causing less deadly repercussions. All of the above factors caused a huge number of deaths in Haiti.

2011 Joplin, Missouri, Tornado

The 2011 Joplin, Missouri, USA, tornado alone killed 161 people as it passed through a densely populated section of this town of slightly over 50,000 people. One police officer also died the day after the tornado, but was not included in the total number of fatalities caused by the event because the officer was struck by lightning while assisting with tornado recovery and cleanup efforts (Paul and Stimers 2014). However, this EF-5 tornado stands as the deadliest single tornado to hit the United States since modern record-keeping began in 1950, surpassing the tornado of June 8, 1953 that killed 116 people in Flint, Michigan (Mustain 2011).[4] No single tornado from 1980 to 2010 has killed more than 40 people (Simmons and Sutter 2012). Over that time period, tornado fatalities in the United States have averaged around 55 per year; yet the Joplin tornado alone killed nearly three times that average

(Simmons and Sutter 2011). To add perspective, the 1925 Tri-State tornado, with a 291-mile (468 km) path, killed 2.4 people per mile; the Joplin tornado caused 27 deaths per mile (Paul and Stimers 2014).

The Storm Prediction Center (SPC) of the National Oceanic and Atmospheric Administration (NOOA) reported that only 45 tornado fatalities occurred in the country during all of 2010, and just 21 in 2009. Average annual tornado deaths in the United States between 2000 and 2010 were 55, but if the 548 deaths caused by tornadoes in 2011 are added, this figure increases to 96 (SPC 2012). Still, the record number of deaths caused by the single tornado in Joplin was far higher than the average annual number of tornado deaths in the country during either time period (i.e., 2000–2010 or 2000–2011). Notably, this occurred when deaths from tornadoes in the U.S. have decreased dramatically over the past century. Calculations from the SPC data indicate that the national tornado fatality rate fell from 1.8 per million in 1925 to 0.06 per million in 2017 (also see Paul and Stimers 2014).

Using questionnaire surveys administered among tornado survivors and informal discussions with Joplin emergency management personnel, local residents, city officials, and others, Paul and Stimers (2012) identified five reasons for the high number of fatalities caused by the 2011 Joplin tornado: (1) the sheer magnitude of the event; (2) its path through commercial and densely populated residential areas of the city; (3) the massive size of area affected by the event; (4) the tangible attributes of affected homes; and (5) the fact that a few Joplin residents disregarded tornado warnings.

The length of the 2011 Joplin tornado track or path was 22.1 miles (35.6 km), which is not considered very long. Typically, tornado paths in the United States range from less than a mile up to 100 mi (150 km). The number of deaths does not only depend on the path's length, but it also depends on whether the tornado path passes over highly populated areas. In that case, deaths are likely to be higher than if it passes over less populated areas. Tornadoes usually spend most or all of their duration over sparsely populated and/or unpopulated areas (Wurman et al. 2007); only about 10 percent of all tornadoes pass through populated areas and almost all of these events miss commercial areas (Stimers 2011). However, of the total track, the Joplin tornado traveled at least a 6-mile (9-km) long path across a densely populated part of the city with winds of more than 200 miles/hour (320 km/hour) (NWS 2011). Over no fewer than 4 (6 km) of those 6 miles (9 km), the tornado was rated EF-5, the first EF-5 tornado in Missouri since the Ruskin Heights tornado struck south of Kansas City in 1957. It also marks the first EF-5 tornado on record in southwest Missouri (Paul and Stimers 2014).

The damage zone, the area across the tornado path or width of the tornado track that sustains tornado damage, covered 7.44 square miles (19.3 square km), equivalent to nearly a quarter of Joplin, which is 31.54 square miles (81.7 square km). According to the 2010 Population Census of the United States, 13,547 people or 27 percent of the city's population resided in the 500 census blocks directly affected by the tornado (U.S. Census Bureau 2012).

The area of damage zone in Joplin was the highest on record in the country. As noted, damage area was approximately 7.44 square miles (19.3 square km), more than seven times the average tornado damage area in the United States, which is less than 1 square mile (2.9 square km) (Simmons and Sutter 2011; Stimers and Paul 2017).

The lack of basements in residential and other structures in Joplin most likely contributed greatly to the high death toll, although to what degree remains uncertain. These structures failed to protect many residents from a high magnitude tornado. Moreover, the city authority provided the tornado warnings 17 minutes before touchdown, which was greater than the national average warning time of approximately 14 minutes (Stimers and Paul 2017). Many Joplin residents also took shelter after receiving the tornado warnings in the most appropriate location (e.g., interior rooms, hallways, bathrooms, closets, or crawl spaces) within their permanent homes or other structures, but a considerable number of these residents died because of lack of a basement. These residents found themselves in situations that were not survivable in an EF-5 tornado. It is unclear to what degree the lack of basements contributed to tornado mortality in Joplin (NWS 2011).

Given the physiography of the region, most notably the hard near-surface rock stratum underlying Joplin, below ground shelters were not common in the city. According to the Jasper County Assessor's Office, nearly 78 percent of houses in the county lack basements, due in part to the rocky ground and high water table (Ryan 2011). Joplin has an even lower percentage of basements than Jasper county communities as a whole. Additionally, 80 percent of the houses in Joplin were older than 40 years in 2011. In 2009, the median house value in Joplin ($93,108) was 34 percent below the Missouri state average of $139,700 (City-Data.com 2011). Older houses were constructed according to the standards of the time, which were far less stringent than today's more rigorous building codes. Many of these older houses are not secured to their foundation; some do not even have a foundation (Paul and Stimers 2012).

Tornado fatalities also depend on the extent of human compliance with tornado warnings, which may correlate with higher instances of deaths. In one of their studies related to the 2011 Joplin tornado, Paul and Stimers (2012) claim that 29 or 23 percent of 126 respondents did not comply with warnings after receiving them. This percentage of non-compliance was higher than the national average primarily because some Joplin residents either had no faith in the tornado warnings or they did not take the warning seriously. In Joplin, the city policy is to activate sirens either when a tornado is reported to be moving toward the city or severe thunderstorm winds are expected to exceed 75 miles per hours (121 km per hour). This policy increased the number of false alarms, and frequent false alarms reduce the effectiveness of hazard warnings (Paul et al. 2015).

Historically, tornado activity around the Joplin area has been not only well above the state average, but also 161 percent greater than the national average. This implies that tornadoes are not rare events for residents of this area. However, historically, most tornados occurring in the Joplin area were of low magnitude, ranging from EF-0 to EF-2. Besides, tornados in this area have been mostly short-track events, and usually displayed fast ground speed. Tornados with these physical characteristics typically do not become deadly. For these reasons, some people in Joplin did not take the severe weather warnings seriously (Paul and Stimers 2012).

Surprisingly, several residents and city officials of Joplin considered that the number of deaths caused by the May 22, 2011 tornado could have been much higher than 161 for several reasons. First, the Joplin High School graduation was held at the time of the tornado passing through the city. The number of deaths would have been higher if the graduation had been held at the school auditorium. Instead, it was held at the Leggett & Platt Athletic Center on the Missouri Southern State University campus. While the Center was not in the damage path, the high school was destroyed. Second, while risking their own lives, the staff of St. John's Regional Medical Center successfully moved 180 patients to safety immediately before the tornado hit the medical facility. The loss of life would have been far greater had they thought only of themselves and not engaged in this heroic action (also see Joplin Globe 2011). Thirdly, the tornado killed three people at the Elks Lodge, which had been preparing for bingo night when the tornado struck. If this tornado had arrived two hours later, there would have been as many as 40 or 50 people in the lodge, and many of them likely would have been killed.

Director of Jasper County Emergency Operations Center (JCEOC), Keith Stammer, maintained that the community of Joplin, to some extent, is lucky in the sense that the tornado occurred on a weekend. He claimed that the death toll would have been much greater if it had occurred on a weekday when the Joplin population reaches an estimated 275,000. This is because of two large medical facilities, a growing medical spin-off industry, numerous restaurants, and a very active trucking industry, as well as shopping facilities that employ many people, many of whom live in the neighboring communities. Joplin also attracts visitors from neighboring communities who seek medical treatments and other services it provides. Indeed, nearly one-quarter of the total deaths were people from outside the damage zone visiting or shopping in parts of Joplin affected by the tornado (Paul and Stimers 2014).

1889 Johnstown flood, Pennsylvania

The Johnstown flood of 1889 was one of the worst catastrophes in the United States, killing 2,209 (McCullough 1968). The event accounted for the largest loss of life in the country from any natural disaster at the time. Still today it is ranked 51 in the deadliest floods in the world. Although number of the average yearly flood fatalities in the United States is not available for the last two centuries, this one is included because of the sheer number of deaths it caused. To provide some comparison, flooding killed 92 people in the United States in 2019, and 182 people in 2018. According to the National Weather Service, 93 and 85 people on average were killed annually for the 2008–2017 and 1988–2017 time periods, respectively (NWS 2020).

The flood begun on May 31, 1889, following several days of heavy rain, which caused the collapse of the South Fork Dam near Johnstown, Pennsylvania. The dam was constructed on the Little Conemaugh River and this dam created a reservoir called Conemaugh Lake, a body of water over two miles (3 km) long. The water in the lake overflowed the dam (owned by the South Fork Fishing and Hunting Club, whose members included industrialists Andrew Carnegie and Henry Clay Frick), and a large section of the dam washed away because of the tremendous pressure exerted by rising water levels (Curtis and Mills 2009). To compound the problem, the dam's spillways were clogged by a large amount of floating debris. Within 15 minutes, the lake was empty and its water rushed from the lake down the valley toward Johnstown at 40 miles (60 km) per hour, destroying the towns of Johnstown and South Fork was destroyed by the floods. It was the first major disaster in which the American Red Cross was involved (McCullough 1968).

Among the deaths, 99 families lost all their members. One-third of the dead were never identified, and bodies were found as far away as Cincinnati (600 miles or 900 km) and as late as 1911. In the affected area, the flood killed about one person out of every 10, and one out of nine in Johnstown itself (McCullough 1968). Several reasons were behind so many deaths. First, the South Fork Dam collapsed because it had suffered many years of neglect and it needed urgent repair by 1889. In this context, the physical condition of the dam can be compared with that of the levee system along the Mississippi River in New Orleans prior to Hurricane Katrina's landfall. Additionally, since its completion, the reservoir had manifested several leakages around the main drain pipes (Ebert 1993).

A telegraphed warning message of the imminent flood came to Johnstown a few minutes before the event. However, flooding is recurrent in the adjacent valley, and no one took the warning seriously because such warnings had proved false many times in the past (McCullough 1968). Thus, there is a striking parallel with regard to faith in hazard warnings for both the 2011 tornado in Joplin and the Johnstown flood. Ultimately, the mountain of water, estimated at 30–40 feet (9–12 m) high and nearly half a mile wide, reached Johnstown with terrific force, carrying debris, boulders, trees, houses, barns, dead farm animals, and massive machinery. People in the path of waters were often crushed as their homes and other structures were swept away. Debris piled 40 feet (12 m) high at the old Stone Bridge in downtown Johnstown, and caught fire, which lasted for three days. Here, 80 people burned to death. Additionally, 2,000 people died in the city due to drowning. Flood deaths were also reported in other cities located between the dam and Johnstown.

1970 The Great Bhola Cyclone, Bangladesh

The Bay of Bengal, which forms Bangladesh's coastline, is one the world's most active areas for tropical cyclones. Although cyclones originating in the Bay of Bengal constitute only less than 10 percent of the global total, they account for 80–90 percent of cyclone deaths in the world (Paul 2009). Fifty-three percent of all the cyclones that have claimed more than 5,000 lives have made landfall on the Bangladesh coast (GoB 2008). Over the last 49 years (1970–2018), on average, about 13,410 people have been killed annually by tropical cyclones in Bangladesh. Two of these cyclones (Bhola Cyclone and Cyclone Gorky) each killed more than 100,000 people, and therefore they are included here.

Thus, a deadly Category 3 cyclone hit former East Pakistan (present day Bangladesh) on the evening of November 12, 1970 and is known as The Great Bhola Cyclone.[5] Recently, the World Meteorological Organization (WMO) officially declared this cyclone the world's all-time deadliest weather event (Salley 2017). Although the death tolls differ by sources, it killed anywhere from 300,000 to 600,000 people (Haque et al. 2012).

Many people died from the Bhola Cyclone largely because of a huge storm surge that overwhelmed the offshore islands and tidal flats along the shores of the Bay of Bengal. The surge was accompanied by torrential downpours and unrelenting winds. The precipitation and winds, combined with a high tide and relatively flat topography, resulted in 33-foot (10 m) storm surges that rank Bhola as the deadliest cyclone in recorded history. Another reason for so many deaths was that the cyclone hit at a time when coastal residents were preparing or eating dinner. In absence of modern predictive equipment, little was known then about cyclones ahead of landfall.

Using remotely sensed satellite imagery, the United States identified the storm's formation in the Bay of Bengal and right way informed the Pakistan government about the storm's formation and location (Penna and Rivers 2013). However, the government started to broadcast cyclone warnings stating moderate intensity on national radio only two to three hours prior to the landfall. But many destitute coastal residents did not own a radio, while the few who had, did not trust the warning and weather reporting due to grossly inaccurate reporting historically (Frank and Husain 1971). Naturally, the bamboo and thatched houses, which were common in the coastal areas at that time, were not strong enough to withstand the cyclone's wind speed and storm surge force. In fact, no public cyclone shelter, which could have saved many lives, existed prior to the 1970 event. Considering the huge loss of life, the Bangladesh government began to build public cyclone shelters for coastal residents in 1972. The shelters are generally two- to three-storied, reinforced concrete buildings that can accommodate more than 1,000 people. The ground floor in each shelter is open and the many shelters serve multiple purposes, particularly as schools, community centers, and government offices (Paul and Dutt 2010).

1991 Cyclone Gorky, Bangladesh

The second deadliest cyclone, a Category 4 storm named Gorky, struck the southeast coast of Bangladesh, north of Chittagong, on the night of April 29, 1991, killing nearly 140,000 people (Haque and Blair 1992). The cyclone was accompanied by tidal surges up to 30 feet (9.1 m) high. It affected 5 million people in eight of Bangladesh's 64 districts (Chowdhury et al. 1993).[6] Many deaths were due to Gorky's physical parameters, problems with the cyclone warning system and poor compliance with such warning, and the inadequate number of public cyclone shelter in the affected areas. Storm surges associated with Cyclone Gorky continuously battered the affected coast for three to four hours. Additionally, the height of the storm surge is considered an important determinant of the number of fatalities caused by a cyclone. Indeed, the height and landfall of a cyclone during high tide are directly associated. Cyclone Gorky made landfall at high tide in the dead of night – another possible reason for the significantly greater fatalities.

Although a warning system was introduced after the Bhola Cyclone in 1970, still it was not well developed and did not function consistently in 1991. Prior to landfall, more warnings had been received by the residents of the offshore islands than in the coastal regions where the overwhelming majority of the 5 million affected people lived. A study reported that only about 60 percent of the residents heard a warning in the coastal areas of Chokoria and Bashkhali; both were devastated by Cyclone Gorky (Haque 1995). Most of those residents heard from the volunteers of the Cyclone Preparedness Program (CPP), which was established in 1972 and stationed in coastal districts to disseminate cyclone warnings among villagers. Their number, 1,991, was inadequate compared to the number of people they served. Another source of warning at that time was radio, but only 16 percent of the households in the Gorky-affected areas owned a functioning radio (Chowdhury et al. 1993).

Surprisingly, only 15 percent of all coastal residents complied with the cyclone warning and took refuge in public shelters. The principal reason for this low rate was lack of trust in the message arising out of repeated false cyclone warnings along coastal Bangladesh. This was largely responsible for non-compliance with evacuation orders during the 1991 cyclone event (Haque 1995; Paul et al. 2010; Ahsan et al. 2016). An additional 16 percent took shelter in neighbors' houses or in public buildings such as mosques, schools and colleges, and government office buildings, which were perceived to be structurally stronger than their residences (Chowdhury et al. 1993).

Another problem associated with the warning system was confusion about its type and number. There were two types of cyclone warning signals: one for maritime or sea ports, and the other for inland river ports. For maritime ports, 11 individual signals are used in different stages of a disaster, and for inland river ports, four separate signals are used. These two types of signal numbers for the same cyclone are confusing to local people.

Note that the increased signal numbers for the sea ports are not based on cyclone intensity. For example, danger signal numbers 5, 6, and 7 are for cyclones with same intensity. Similarly, great danger signals 8, 9, and 10 are for cyclones with same intensity, whereas signal number 11 means communications have broken down with the meteorological warning center. Twenty-four hours before landfall, Cyclone Gorky changed course, and the Bangladesh authority reduced the signal number from 10 to nine, which created a false impression among the coastal residents that the cyclone had weakened (Chowdhury et al. 1993).

Additionally, cyclone shelters were few in relation to need of the people affected by Cyclone Gorky. Moreover, the shelters suffered many problems such as lack of drinking water and latrines, poor maintenance, and no separate room for females. According to traditional Bangladesh culture, males and females should not be in the same room. Other reasons for not going to a shelter were long distance, failure to realize danger, lack of time, feeling safe at home, and fear of burglary. Still, Chowdhury et al. (1993) claimed that at least 20 percent more deaths would have occurred in the absence of the available shelters. Poor housing, which is closely linked with widespread poverty, is another reason for too many deaths caused by the 1991 cyclone. Only 3 percent of the houses were strong enough to withstand the onslaught of Gorky-induced tidal surges. This is because most houses in coastal zones of Bangladesh were made of straw and mud (Chowdhury et al. 1993).

2010 Cyclone Nargis, Myanmar

On the evening of May 2, 2008, Cyclone Nargis made landfall in the Ayeyarwady (Irrawaddy) delta region in southern Myanmar. With winds up to 150 miles/hour (240 km/hour), this Category 4 cyclone was the worst natural disaster in the country's recorded history. Its sizable storm surge height of 13 feet (4 m) flooded 25 miles (40 km) inland and killed nearly 138,000 people, and another 53,800 went missing. A total of 37 townships were affected by the cyclone, and overall 2.4 million people were severely affected. Cyclone Nargis was the worst to hit Asia since 1991, when 131,539 people died from Cyclone Gorky in Bangladesh. In terms of the number of deaths, it was the worst natural disaster to hit Asia since the 2004 tsunami that killed at least 226,400 people (Reuters 2009).

The principal cause of the staggering number of deaths was that the Myanmar military government was not prepared for this disaster. Although the government claimed it had warned people about the storm, critics contend the junta did not issue a cyclone warning and failed to evacuate a single person from the potential affected areas, even though information was available about the expected time and location of landfall. Moreover, the government did not have a radar network to help predict the location and height of the tidal surge. Still, some experts maintain that if the government had been able to provide early warning of the cyclone, residents of low-lying areas would have had no place to take refuge in the absence of sturdy cyclone shelters (AP 2008). Ultimately, many of the deaths occurred due to the storm surge that hit the country's low-lying but also very populous Irrawaddy Delta region.

2013 Typhoon Haiyan, the Philippines

Typhoon Haiyan, known in the Philippines as Super Typhoon Yolanda, ravaged the central coast of the Philippines as a Category 5 storm in the early morning of Friday, November 8, 2013. With estimated sustained winds of 195 miles/hour (315 km/hour), this was the second deadliest typhoon in the country since 1867. It killed 6,300 people, causing 28,688 injuries and 1,062 to go missing. The death toll was relatively lower than for the Bhola Cyclone, Cyclone Gorky, and Cyclone Nargis, but it is included for three reasons: its massive loss of life in the Philippines, the fact that most of the deaths could have been prevented, and because after the typhoon, President Benigno Aquino III called for an investigation into why so many people died. Ultimately, Typhoon Haiyan affected more than 14 million people across 44 provinces in the country, displacing 4.1 million people, and disrupting the livelihoods of 5.9 million workers. The typhoon also affected other Southeast Asia countries such as Vietnam, China, and Taiwan (CRS 2014).

Typhoon Haiyan destructiveness was due to a combination of meteorology, geography, population density, and poverty factors. It made landfall in Tacloban, a coastal port city of San Pablo Bay in Leyte province. With 220,000 inhabitants, the city is located at the tip of a funnel-shaped bay in the Leyte Gulf. Tacloban experienced storm surges 16 feet (5 m) high, and the enclosed nature of the bay magnified the storm surges. Much of the city and two towns to its south, Palo and Tanauan, were inundated because their elevation is less than the height of the storm surge. The location of the city along Haiyan's track led to Tacloban receiving the brunt of the storm's wind, which destroyed up to 80 percent of the city's buildings, causing many deaths of those were unable to get to designated shelters (Mersereau 2013).

Typhoon Haiyan not only affected areas of high population density (2,900 persons/square miles or 1,160 persons/square km), but also some of the poorest parts of the Philippines, where people are either farmers or fishers, and live in houses that can barely withstand strong wind. With the storm's landfall anticipated, government officials in Tacloban and Manila began to prepare for the typhoon. The city officials opened 23 evacuation centers, of which 22 were school buildings. Although no mandatory evacuation order was issued, hundreds of officials went door-to-door to encourage residents to take shelter in an evacuation center. Because of the mountainous topography, moving people out of Tacloban into other areas was not possible.

Also, some of the designated evacuation centers could not withstand the storm's wrath and collapsed, causing many deaths. The metal sheet roofs of most of the designated shelters were not tightly tied to the walls, and because of this, several shelter were blown away by strong wind. Many shelters were also not on elevated land, and thus were liable to inundation from storm surges. In fact, of 300 people who sought safety in one of the shelters in Tacloban, at least 23 died by drowning when storm surge water inundated the shelter (CRS 2014; McPherson et al. 2015).

Moreover, the evacuation rate was very low; only about 7 percent of the population of Tacloban was evacuated (Yamada 2017). This rate was even lower among vulnerable residents, and many of them opted to stay in their flimsy homes. Three reasons were responsible for the low rate of evacuation. First, residents were reluctant to leave their home for fear of theft. Many others were doubtful about the danger of impending disaster. Finally, when a tsunami warning was issued in February of 2013, the government initiated mass evacuation and it turned out to be a false alarm. Thus, in the case of Typhoon Haiyan, the government officials did not initiate mass evacuation. Another reason for low rate of evacuation was that the typhoon warning mentioned the term "surge," which was unfamiliar even to those who had lived for years with fierce storms. Thus, the term's implications were not understood by the populace (Yamada 2017). Rather, they were familiar with the term "tsunami," which is associated with earthquakes in the sea, but not with the term tropical storms.

2004 Indian Ocean Tsunami

On the morning of December 26, 2004, a 9.0-magnitude earthquake occurred in the ocean off the western coast of North Sumatra, Indonesia. This event quickly turned into the worst disaster in a century, resulting in a tsunami that crashed into the coasts of 14 countries. The tsunami killed 226,400 from over 50 countries, including 9,000 foreign tourists, who came mostly from European countries largely to Thailand (Paul 2019). Among the four most affected countries, with 167,540 deaths, Indonesia ranked at the top in terms of tsunami fatalities, followed by Sri Lanka (35,322), India (16,269), and Thailand (8,212) (Karan 2011).

In Indonesia, Aceh was the worst affected region because of the proximity to the epicenter of the earthquake. This province has been under martial law since the early 1990s because the military wing of the Free Aceh Movement has been fighting for separation from Indonesia. Similarly, Tamil Tiger rebels in Sri Lanka had been continuing a separatist movement for more than two decades and had established a parallel administration in the northeast part of the country. Like Aceh, this region was severely ravaged by the tsunami (Paul 2013). Apart from complex emergencies in these two parts of the two most affected countries, many tsunami deaths were also caused by absence of preparedness, lack of pre-disaster tsunami warnings, and scale of impact.

Tsunami waves reached the coasts of affected countries in Southeast Asia multiple times. The first wave came and then retreated and was followed by at least three more. After the first wave retreated, coastal residents returned to look for their relatives, friends, and neighbors, and some of them died in the second wave. Also, people who were injured in the initial wave may have been more severely injured or killed in subsequent waves (Yamada et al. 2006). In parts of Sri Lanka, a wave over three feet high (1 m) struck initially, followed 10 minutes later by a second wave that was over 30 feet (10 m) high (Yamada et al. 2006). Thus, a considerable number of Sri Lanka's tsunami deaths are attributable to subsequent tsunami waves. Additionally, of the total 19 million people, one to two million were directly affected (5–10 percent) by the 2004 Indian Ocean Tsunami. Being largely distant from the source of the origin of the tsunami, nevertheless, Sri Lanka recorded the second largest number of deaths among the affected countries due to multiple waves and exposure to a relatively large part of the population.

Another reason is also associated with a very large number of deaths caused by the 2004 Indian Ocean Tsunami. Available information suggests that up to four times as many women as men died in this extreme event. In some coastal villages in Indonesia and Thailand, women accounted for up to 80 percent of those killed. The reason for this disproportionate gender impact was that women were waiting near the coast for their returning husbands to sort out catches. Generally, husbands catch fish in the ocean at night and return to the coast in the morning. If women were not helping sort fish every morning, or tsunami occurred other than in the morning, far fewer deaths could occur. Thus, timing is an important factor in the number of deaths caused by natural disasters.

Relatively few deaths

1960 Valdivia earthquake, Chile

The largest recorded earthquake in the twentieth century struck Chile on May 22, 1960, at 3:11 pm. The epicenter of the 9.5 magnitude earthquake was approximately 100 miles (160 km) off the coast of southern Chile, parallel to the city of Valdivia. For this reason, the event is called the Valdivia earthquake. The earthquake occurred at a depth of 21 miles (33 km) and generated the largest tsunami in the Pacific region for at least 500 years. The tsunami waves reached a height of 82 feet (25 m), and were destructive not only along the coast of Chile but also across the Pacific in Hawaii, Japan, and the Philippines.

However, the world's largest earthquake and tsunami on record did not cause the greatest death toll. The estimated fatalities range between 490 and 5,700. This is because a large foreshock that struck only 30 minutes before the main event sent many people from their homes to the street. It saved them from the main shock, and their abandoned houses ultimately were destroyed during the main shock (Brumbaugh 1999).[7] In fact, before the major shock, three foreshocks provided early warning of the major earthquake. The first foreshock, which measured 8.3 magnitude on the Richter scale and occurred less than 24 hours before the major shock, severely damaged the coastal town of Concepción (BSL 2015). The second and third foreshocks occurred on May 22, and measured 7.1 and 7.5 on the Richter scale, respectively. The last two foreshocks did not cause any fatality. These three successive foreshocks are together called the 1960 Concepcion earthquakes (Pallardy n.d.).

2010 Chile earthquake

In 2010, two major earthquakes shook the Americas. The first of these two events hit Haiti with a magnitude of 7.0 on the Richter scale and killed at least 222,750 people. The second earthquake, which struck Chile one month after the Haiti earthquake on February 27, was 500 times stronger – 8.8 on the Richter scale, yet it killed only 521. Despite much higher magnitude and the event caused tsunami waves in Chilean coast as high as 50 feet (15 m), far fewer people died than did in the Haiti earthquake. This was primarily because of the country being well prepared for the event. For example, many people were evacuated from the central part of the Pacific coast prior to the tsunami. Many more participated in annual evacuation drills prior to the 2010 earthquake. Thus, the combination of early warning system and preparation were credited with saving lives in the case of Chile, but Haiti was not prepared at all given its low risk of earthquakes and the fact that Haitians are preoccupied with the threat of hurricanes (BSL 2015).

These conclusions indicate that the level of economic development seems unrelated to the number of deaths caused by earthquake events in Haiti and Chile. However, while it was stronger, the Chilean earthquake also occurred 22 miles (35.5 km) below the earth's surface, it was twice as deep as Haiti's. There is another reason why buildings in Chile withstood the earthquake better than in Haiti: improved earthquake detection and construction improvement. The latter country had not suffered an earthquake of this magnitude since the eighteenth century, while Chile experienced nearly two dozen major earthquakes in the twentieth century. In fact, during the past 150 years or so, Chile has had more giant earthquakes with magnitudes of eight and larger than any other country in the world (BSL 2015).

Because of the region's violent tectonic history, GPS sensors had been installed across Chile and neighboring countries to monitor and detect seismic activities. Also, after the devastating 1960 earthquake of 9.5 magnitude, Chile introduced stringent building codes, which were revised many times during the 1990s. This limited the collapse of buildings resulting in fewer deaths due to failure of such structures. In addition, the country's infrastructure remained intact, which facilitated quick recovery and response. In fact, 150 of the 521 deaths were caused by the tsunami, so the earthquake directly caused nearly 400 deaths. The Chilean government dispatched more than 10,000 troops to the devastated areas to distribute emergency assistance and restored power lines within a day of the earthquake. Shortly after, reconstruction efforts were under way. Quick, adequate, and coordinated response also contributed overall to the few casualties of the 2010 earthquake in Chile (Pallardy and Rafferty n.d.).

According to the American Red Cross Multidisciplinary Team that visited Chile in July 2010 in an effort to reduce future earthquake losses for California, several factors were responsible for the relatively few deaths. First, Chile has strict building codes, which was the prime reason for the strong performance of the built environment. The country has a law that holds building owners accountable for any loss in a building from inadequate compliance with code for the first 10 years of a building's existence. A second factor was the limited number of fires after the earthquake. Third, in many affected areas, the local emergency response was very effective in that there was close coordination between emergency management, fire, and police agencies. In short, local responders were empowered to respond without communication with the central government. Overall, a high level of knowledge about earthquakes and tsunamis by most residents of Chile was the fourth factor. This helped them respond more appropriately after the earthquake (USGS and American Red Cross 2011).

John Fetter (2017) also added one more reason: wealth. "Chilean citizens enjoy an average income 10 times that of Haitians. Fully 80 percent of Haitians live below the poverty line, compared to fewer than 20 percent of Chileans" (Fetter 2017). With more wealth and resources, Chile invested relatively more than Haiti in building sound infrastructure, providing decent services, and ensuring proper regulations.

Two more reasons support the much lower death rate in Chile. The former earthquake occurred in the ocean in the subduction zone, which was associated with tsunami. The Haiti earthquake, on the other hand, originated on land and thus did not cause tsunami. Geologists believe that in general earthquakes that originate in the ocean cause less damage than do earthquakes originating on land. Moreover, offshore earthquakes generate less ground shaking than onshore earthquakes because the intensity of ground shaking diminishes with distance (Wayman 2010). The last reason was that the Chilean earthquake occurred 70 miles away from the city of Concepcion (with population 200,000), while the Haitian earthquake originated only 15.5 miles (25 km) southwest of the city center of Port-au-Prince with 2.5 million in its metropolitan area. Thus, the differences with respect to physical characteristics of the two earthquakes, geological differences, and social factors explain why the larger earthquake in Chile was much less deadly (Table 2.2).[8]

2007 Cyclone Sidr, Bangladesh

A Category 4 storm, named Cyclone Sidr, struck the southwestern coast of Bangladesh on the night of November 15, 2007, killing fewer than 4,000 people (Paul 2009). Considering historically deadly tropical cyclones in

Table 2.2 Comparison of two earthquakes struck Haiti and Chile in 2010: selected parameters

Parameter	*Haiti*	*Chile*
Date and time	January 12 at 4:53 pm	February 22 at 3:34 am
Magnitude (on Richter scale)	7.0	8.8
Duration	15–30 seconds	3 minutes
Epicenter depth (miles/km)	13/20	22/35
Epicenter location	On land	On sea
Preparedness	Nil	Well
Population center	Close	Far
Tsunami	No	Yes
Number of deaths	222,750	521 including tsunami

Sources: Compiles from different sources.

Bangladesh, this figure was strikingly lower than anticipated and widely attributed to several factors. First, the vastly improved emergency management system developed after the Cyclone Gorky was one of the most significant factors that minimized the loss of lives from Sidr. This is reflected in timely cyclone warnings and aggressive initiatives by emergency management authorities for the successful evacuation of coastal residents from the projected cyclone path (Shamsuddoha and Chowdhury 2007; Blake 2008; Hossain et al. 2008). Those residents were evacuated to an expanding network of public cyclone shelters. While official sources claimed that between 33 percent and 40 percent of coastal residents were evacuated to public shelters ahead of the landfall of Sidr, the rate of evacuation was only 15 percent in the case of Cyclone Gorky (Robinson 2007; GoB 2008; Paul et al. 2010).

As indicated, timely cyclone warnings and evacuation were possible for Cyclone Sidr because remarkable progress had been made following Cyclone Gorky to improve the early warning systems and implement several other cyclone preparedness and mitigation measures. For example, the number of community-based volunteers in the CPP more than doubled between 1991 and 2007. In 1991, there were 20,000 CPP volunteers stationed in 17 coastal districts (Chowdhury et al. 1993). That number increased to 42,675 prior to the landfall of Sidr (Paul 2009). Using microphones, megaphones and door-to-door contact, CPP volunteers warned the public about the cyclone and encouraged them to evacuate prior to landfall. Moreover, CPP volunteers were joined by volunteers from Red Crescent Society, local government officials, workers from Nongovernmental Organizations (NGOs), and some villagers (Paul 2009).

Cyclone preparedness in Bangladesh with regards to the number of public cyclone centers also markedly improved between the landfalls of Cyclone Gorky and Cyclone Sidr. In 1992, there were 512 public cyclone shelters on coastal Bangladesh (Ikeda 1995), which had increased to 3,970 by the time of landfall of Cyclone Sidr (Shamsuddoha and Chowdhury 2007).[9] Increasing the number of CPP volunteers and shelters had a positive impact on the residents of the southwestern coast, who received early cyclone warnings and on their evacuation rate. Based on field data, Paul (2009) reported that about 86 percent of all households surveyed were aware of the cyclone warning and evacuation order in advance of Sidr's landfall, and Robinson (2007) maintained that some 3.2 million of 8.0 million coastal residents (40 percent) evacuated their home prior to the arrival of the cyclone.

Much lower than expected fatalities were also the result of a number of physical characteristics of Cyclone Sidr, such as landfall time, site of landfall, duration of the storm and storm surge, varied coastal ecology, and coastal embankment. Cyclone Sidr struck the southeastern coast in early night, which provided residents of the affected areas time to look for suitable shelter. Additionally, storm and storm surge lasted less than three hours, and affected areas experienced only one storm surge instead of a succession of surges as was the case of Cyclone Gorky. In fact, Cyclone Gorky surges brought walls of water directly from the sea, while Cyclone Sidr storm surge water entered settlements mostly from coastal estuary rivers, reducing the force of surge water. Timing, duration, and the number of surges were factors responsible for the surprisingly low number of fatalities (Paul 2009).

The fourth reason is that unlike Cyclone Gorky, Cyclone Sidr made landfall during low tide, which reduced the surge height and hence its impact on the number of fatalities. The cyclone made landfall on the southeastern part of the Sundarbans between Hiron Point and the Baleswar River.[10] It is the site of the world's largest mangrove forest, lying along the southwestern coast, and thus it acted as a buffer zone, reducing the effects of Sidr by slowing down and lowering the deadly storm surge levels. The thick growth of mangrove trees also successfully reduced the wind speed so that most villages behind the trees were saved. In fact, the southern part of the Sundarbans is virtually unpopulated, while population density in the rest of the forest is much lower than the national average. Even so, the southeastern coast of Bangladesh has a higher population density than the southwestern and southcentral coasts of Bangladesh, indicating that the low population density of the southwestern coast minimized the loss of lives from Sidr (Paul 2009).

The fifth reason for the lowered death toll from Cyclone Sidr is the varied geomorphology of Bangladesh's coast. The 500-mile (800 km) long coast is divided into three ecological and geographical zones: the southwest, which is dominated by the mangrove forest of the Sundarbans; the Meghna estuary and the vast active delta zone in the central portion of the country; and the eastern zone, which has a narrow, straight-line coast running parallel to the geologically young (Tertiary) folded hill ranges (Rashid and Paul 2014). Unlike the other two coastal zones, there are no large islands off the southwestern coast, such as Bhola, Hatiya, and Sandwip, which are heavily populated islands. In contrast, islands located south of the southwestern coast are small in size, few in number, and not permanently settled. This seems to be associated with the lower than expected death toll caused by Cyclone Sidr.

The sixth reason is that coastal embankments, which started to be constructed in the early 1960s to protect the coastal area from tidal inundation and save agricultural lands, reduced the number of deaths caused by Cyclone Sidr. There is evidence that many of these embankments provided an effective buffer during the storm surges. In most cases, these embankments were not properly maintained and repaired; still, many of them did not fail when the Sidr storm surge washed over them. Also, the network of embankments was longer in 2007 than in 1991 when Cyclone Gorky battered the southeastern coast (Paul 2009).

A religious factor was the seventh and final reason for many regarding the lower than expected deaths caused by Cyclone Sidr. As in many traditional societies, sensing the "imminent" danger, many coastal residents engaged in prayers, and subsequently, many residents of the affected areas perceived that God had heard and granted their prayers (Paul 2009). Coastal residents found enough time for praying because the Bangladesh government began to broadcast cyclone warnings five days before Cyclone Sidr made landfall (Paul and Dutt 2010).

2009 Cyclone Aila, Bangladesh and India

Cyclone Aila, struck the southwestern coastline of Bangladesh and the eastern coast of the adjacent Indian State of West Bengal on May 25, 2009. It was a Category 1 cyclone, a relatively weak storm that was responsible for the deaths of 327 people in both countries – 190 in Bangladesh and 137 in India (Paul and Chatterjee 2019). Aila made landfall at 6:25 pm local time along the coast of the Indian Sundarbans – not very far from the coast of Bangladesh's Sundarbans. The Sundarbans coast, which contains the world's largest mangrove forest, is home to numerous threatened species, including the famous Royal Bengal tigers. Approximately two-thirds of the forests are located within Bangladesh and the remaining one-third is in West Bengal, India (Rashid and Paul 2014).

Cyclone Aila made landfall at high tide with sustained wind speeds ranging from 46 to 75 mph (74–120 km/h) and resulted in deadly storm surges up to 22 feet (6.7 m) high. Despite landfall at high tide, the cyclone killed fewer people than it might have for several reasons. First, Aila made landfall in the Sundarbans coast, which is less populated, and the cyclone travelled north through the dense mangrove forest, which reduced its destructive impacts. Also, the 5,000-mile (7,500 km) long embankment network built in the 1960s protected coastal residents from Cyclone Aila and associated tidal surges. Second, the early warning dissemination and preparedness helped to reduce human casualty during Aila. Third, a massive evacuation was undertaken in both Bangladesh and India prior to Aila's landfall. For example, the Bangladesh Meteorological Department (BMD) started to provide warnings about Cyclone Aila as early as May 23, 2009, enabling evacuation of 400,000 people in five southwestern coastal districts of the country to public cyclone shelters and safer locations (IFRC 2010).

2013 Cyclone Phailin, Odisha, India

On October 12, 2013, at 9 pm local time, Cyclone Phailin made landfall near Gopalpur, in India's Odisha state. The cyclone formed as a tropical depression on October 4, 2013, near the Andaman Sea very close to Thailand. Accordingly, the Indian government started to evacuate residents from low-lying coastal areas to stormproof structures on October 10, 2013. Cyclone Phailin was the strongest storm to hit the coasts of Odisha and Andhra Pradesh states in 14 years, bringing winds of 140 miles/hour (225 km/hour) along 250 miles (241 km) of the eastern coastlines of India. Indian authorities expected Phailin to come ashore as a super cyclone, with winds in excess of 157 miles per hour (252 km per hour), but the storm made landfall as a Category 4. The cyclone was accompanied by a storm surge of 5 feet (1.5 m) and heavy rainfall that caused extensive floods in the major river basins. Approximately 13 million people of the four states (Andhra Pradesh, Bihar, Chhattisgarh, and Odisha) were affected by this event (Singh and Jeffries 2013).

Cyclone Phailin killed only 45 people, 44 in Odisha and one in Andhra Pradesh. Most of the deaths were caused by falling trees and/or collapsing houses. Compared to the magnitude of the storm, Cyclone Phailin accounted for far fewer deaths than might have occurred primarily because of massive evacuation prior to the cyclone's landfall and early warning of the cyclone by the Indian Meteorological Department (IMD) in Delhi. The warnings started five days before the cyclone's landfall as a red alert, the highest alert from the IMD. The warnings were very effective as they specified where and what type of damage could be expected to houses and infrastructure. Additionally, the IMD strictly monitored Cyclone Phailin's path, speed, and direction.

More than one million people were evacuated from 18 coastal districts, often by force, ahead of its landfall by the Indian and respective state governments to avoid a repeat of the death toll of the 1999 Odisha Cyclone, which killed 10,000 people.[11] This was the largest evacuation for a storm in India's history and was possible because of several factors: the coordination between actors, the availability of core information, effective evacuation planning, flexibility in the standard operating procedures, and responders' dedication and commitment to save lives (Ray-Bennett 2016). Two days prior to the landfall, the Odisha Disaster Management Authority (OSDMA), along with the National Disaster Management Authority (NDMA), started conducting mock drills in the cyclone shelters. These drills were helpful to reduce the impacts of the approaching cyclone. Because the number of cyclone shelters would not accommodate the population that was to be evacuated, the state government opened nearly 250 emergency shelters in sturdy buildings like schools, colleges, community halls, and public buildings (Padhy et al. 2015).

State authorities also stored food and medicine in the potential affected areas before the landfall, and thus reduced the indirect deaths. Another reason for far fewer deaths was that the landfall site was on a steep continental shelf, meaning fewer low-lying areas were vulnerable to storm surge (Samenow 2013). After the deadly Orissa 1999 cyclone, the Indian government adopted a, "zero-casualty approach." Following that cyclone, the IMD modernized its early warning system and enhanced the space technology (Ray-Bennett 2016). In addition, the Odisha government opened 24-hour control rooms in district offices and district officials were equipped with satellite phones to ensure effective functioning of the evacuation. Additionally, the telecom operating service of the Indian Government stored enough fuel to adequately operate the generators of the telephone exchange and the cell towers (Padhy et al. 2015).

Since the 1999 super cyclone, Odisha and the central government initiated several measures to adequately prepare for any future natural disasters. The state government increased interaction among the national government, IMD, and NGO agencies, and the at-risk communities. Also, the Indian government passed the first Disaster Management Act in 2007, which created a knowledge network that included the IMD, the Earth System Science Observation, the Indian Space Research Organization, the Central Water Commission, the Geological Survey of India, and the National Remote Sensing Centre. This network was crucial in generating core information during Phailin (Ray-Bennett 2016). The proactive political leadership, which aligned with the goal of saving lives in disaster preparedness, is also a key reason for the unexpectedly few deaths caused by the 2013 cyclone in Odisha. Thus political commitment can foster a culture of disaster preparedness.

Jason Samenow (2013) added two more reasons for unusually few deaths caused by Cyclone Phailin: the cyclone substantially weakened prior to and during landfall, and the storm's intensity may have been overestimated. As noted, at the landfall site, the maximum sustained winds were around 140 miles/hour (225 km/hour) whereas they have been at least 160 miles/hour (257 km/hour) in the 24 hours preceding. Samenow (2013) provided three possible reasons for the weakened of the storm: (1) its core was reorganizing, (2) the storm was moving slowly enough before landfall to stir up cold water from deep water underneath, and (3) its interaction with land.

2015 Cyclone Pam, Vanuatu

In early March, 2015, Cyclone Pam struck several countries of the South Pacific basin such as the Solomon Islands, New Caledonia, Tuvalu, Fiji, New Zealand, and Vanuatu. Of the more than 80 islands, several major islands of Vanuatu were the hardest hit. Vanuatu is an archipelago that spreads north-south over 867 miles (1,300 km) in the South Pacific and is home to about 270,000 people. With wind gusts of up to 200 miles (320 km) an hour, this Category 5 cyclone made landfall on March 13, 2015, on the southeast corner of Efate Island, where the capital of Vanuatu, Port Vila of, is located. Overall, the storm affected 188,000 people of the country, i.e., more than half of the national population. Approximately 17,000 buildings, about 80 percent of the national housing stock, including schools and clinics, were either damaged or destroyed (GoRV 2015).

Cyclone Pam caused the deaths of 15 people, yet given the magnitude of the cyclone, and that most families lived in rural areas with homes made of wood and mud, the death toll was considered very low. John Handmer and Hannah Iveson (2017) identified four important factors for the low death toll: effective warnings, self-reliance and traditional knowledge and preparation, training and evacuation, and shelter and housing.

One principal reason for low fatality was that Vanuatu was well prepared for the cyclone. As a tropical disturbance, Cyclone Pam formed east of the Solomon Islands on March 6, 2015. On March 8, the disturbance reached tropical depression intensity, and on the next day, it was upgraded to tropical cyclone status when it was 667 miles (1,000 km) northeast of the Solomon Islands. Its movement was relatively slow as it headed toward Vanuatu. Thus, the Vanuatu government had enough time to provide early warning and evacuate people from its path. Although Cyclone Pam made landfall on Efate Island at night, it hit the islands of Tanna and Erromango, south of Efate Island, during subsequent daylight hours. This helped residents of these two islands to monitor the path of the cyclone and take shelter accordingly.

The government of Vanuatu issued initial warnings two days before the cyclone made landfall in the country. Almost all the at-risk population was aware of the approaching cyclone. For the first time, the Vanuatu government used an SMS warning system to alert people of the approaching cyclone. It sent text messages by cell phone, containing condensed versions of warnings from the national meteorology service that included progression, location, and direction of the cyclone; moreover, the messages included the threat level, including color code, and were sent every three hours as the cyclone intensified. As the cyclone came nearer to the land, the messages were sent hourly. Prior to Cyclone Pam between 80 and 90 percent of people had access to mobile phones. Thus, technology played a very useful means of communicating critical information to people.

Prior to the cyclone, aid organizations distributed disaster response materials and kits to Community Disaster Committee (CDC) members (Handmer and Iveson 2017). In addition to official early warnings, people also suspected that a cyclone was approaching by observing nature, such as the abundance of a particular bird in the sky. Accordingly, people prepared for the event by cutting down banana leaves to prevent the trees from falling, and tying down roofs or evacuating to safer buildings. Almost all inland villages have at least one sturdy building for shelter during a cyclone. Also, two days prior to the landfall, Vanuatu Red Cross designated many local churches as evacuation centers.

Furthermore, traditional rural houses in Vanuatu are made of locally available materials such as bamboo, local timber, and leaves and thatch for roof. These building materials are not heavy, and thus withstand the pressure of a cyclone. If the roof collapses, the cyclone wind cannot get under the roof and lift it off. The materials are lashed together rather than nailed together, and that gives the building ability to flex. After Cyclone Pam, a report by the Vanuatu Shelter Cluster claimed that traditional houses survived better than houses with cement sheeting or iron roofing and houses built using traditional materials but in modern styles (Shelter Cluster 2015).

The Red Cross in Vanuatu spent a lot of time training people to prepare for cyclones during the early 2010s. The organization and other NGOs such as Cooperative for Assistance and Relief Everywhere (CARE) and Save the Children taught them techniques to secure houses and store emergency food, interpret cyclone warnings, and develop evacuation plans (Bolitho 2015).

Conclusion

Whether natural disasters cause too many or rather few human deaths depends on many interrelated factors. Despite their diversity, disasters have many common physical dimensions, which influence the number of deaths. These dimensions include magnitude, intensity, duration, and areal or spatial extent. Time of day and day of the week also determine number of deaths, particularly regarding tornadoes, cyclones/hurricanes/typhoons, and earthquakes. For example, the height of storm surges and landfall of a cyclone during high tide are often directly associated with high mortality. As noted, one of the primary reasons for too many deaths caused by Category 4 tropical Cyclone Gorky in Bangladesh was that it made landfall at high tide in the dead of night. Cyclone Aila also made landfall at high tide, but killed only 327 people. However, it was a Category 1 cyclone and made landfall on the Sundarbans coast. Other factors were also responsible for the relatively low death toll caused by Cyclone Aila. These factors have already been discussed in this chapter.

The timing of an extreme event is the most important factor in the number of deaths caused by earthquakes. For example, in 1994, the Northridge, California earthquake occurred at 4:31 am before roads were crowded with commuters. The earthquake killed only 57 people, however, it is suspected that the death toll was probably much lower than would have been had the event occurred hours earlier or later during rush hour when commuters would have been returning to their homes. Similarly, the 1989 Loma Prieta earthquake in northern California killed 63 people. This earthquake occurred at 5:04 pm local time when people were on the roads, returning home from work. Most deaths were caused by collapsing of interstate highways and the crushing of cars on multilevel highways. However, because The World Series baseball game featuring two local teams was scheduled to start at the time of the earthquake's occurrence, fewer people were commuting than would normally have been the case at that time of day (Montz et al. 2017).

Earthquakes kill more people, particularly in developing countries, if they hit during school hours. For example, the 2005 Kashmir earthquake struck Pakistan-administered Kashmir and parts of India, and Afghanistan on October 8, 2005, at local time 8:50 am. This 7.6 magnitude earthquake killed 87,350 people, 50–60 percent of whom were children. The earthquake destroyed 10,000 schools and killed at least 32,000 students and teachers, who were in the school buildings at the time of the earthquake (Ozerdem 2006). If the earthquake had occurred during recess or on a weekly holiday, the death toll would have been much lower. In contrast to the Kashmir earthquake, an earthquake of magnitude 7.8 struck Nepal on April 25, 2015 at noon time. This earthquake occurred on a weekly holiday and killed 8,857 people. If the event had occurred on a week day, deaths, particularly among children, would likely have been higher (Paul and Ramekar 2018). At the time of the earthquake, many children were outdoors instead of inside school buildings. Being indoors or outdoors during earthquakes can affect the number of resulting fatalities because most deaths are the result of damage to buildings or other structures.[12]

Also, epicentral depth has a negative impact on deaths. Specifically, the origin of earthquakes in shallow depths generally mean more deaths compared to if they originate deeper in the earth surface.

Naturally, preparedness and early warning systems are significant determinants of human deaths for all types of natural disasters. For this reason, these two measures are the keys to reducing deaths from any natural disaster, particularly from rapid onset disasters such as earthquakes. Contrary to this, absence of early warning system and lack of emergency preparedness will result in excess deaths both during and after the event. Preparedness is the measures and policies taken by disaster managers before an event occurs that reduce the negative impact, including deaths. Disaster tolls are also the product of the meteorology, geography, population density, extent of poverty, and efficacy of a political system in establishing sound preparation and response procedures.

Notes

1. The Modified Mercalli Intensity (MMI) scale depicts shaking severity and it ranges from I (not felt) through XII (nearly total damage). An earthquake has a single magnitude that indicates the total amount of energy released by the earthquake. However, the amount of shaking experienced at different locations varies based on several factors: magnitude, distance from the epicenter, and surface soil composition. The Italian volcanologist Giuseppe Mercalli originally developed this intensity scale in 1883. Since then it has been modified several times with the latest modification being done by Harry O. Wood and Frank Neumann in 1931.
2. Luckily the earthquake occurred at a time when many people were outside and returning from work or school to home and/or outside playing and talking (Daniell et al. 2013).

3. Had the 2010 Haiti earthquake struck during the night or earlier in the day when more people would have been at school or work, the fatalities likely would have been even greater.
4. Based on estimated wind speeds and related damage, a tornado in the United States was assigned a rating from F-0 to F-5 from 1971 through 2007. The ranking was developed by Theodore Fujita and was called the Fujita scale or F-Scale. The original Fujita Scale was revised in February 2007 to reflect better examinations of tornado damage survey. This revised scale is called the Enhanced Fujita Scale or EF-Scale (Paul 2011).
5. The Saffir-Simpson Hurricane Scale is used to measure both intensity and magnitude of tropical cyclones/hurricanes/typhoons. Based on a five-point scale, it measures strength of hurricanes, and assigns in ascending order Categories 1 (no real damage to building structures) through 5 (heavy damage). The Saffir-Simpson Hurricane Scale is dependent on the wind speed, which, in turn, causes damage to property (Paul 2011).
6. A district is the second largest administrative unit in Bangladesh and contains nearly three million people.
7. Experts believe that over 1,000 deaths in the 1960 Chile earthquake were caused by tsunami.
8. Almost five months after the Nepal earthquake, an earthquake significantly more powerful struck Chile on September 16, 2015, at 19:54:33 local time. The earthquake measured at 8.3 in the Richter scale lasted between three and five minutes. The epicenter was near populated areas, just 175 miles (282 km) north of the capital Santiago. It caused deaths of only 11 people and damaged only a few hundred houses (Vijaykumar 2015). This unusually low number of deaths was for three reasons. After the 1960 earthquake, Chile established strict anti-seismic building codes and modernized its tsunami warning system. Finally, the Chilean government was able to evacuate more than one million people from coastal areas in a matter of hours, escaping the tsunami waves, some of which were 15 feet (4.6 m) high in the region of Coquimbo (Vijaykumar 2015).
9. This translates to annual increase of cyclone shelter for the period was nearly 49 percent, while the population grew at a rate of 2.2 percent per year in the coastal districts (Rashid and Paul 2014).
10. Hiron Point is a protected wildlife sanctuary in the south of the Sundarbans and an important tourist spot in Bangladesh. As a wildlife sanctuary, this place is home to Royal Bengal Tigers, Chitra deer, wild pigs, monkeys, birds, and reptiles. Hiron Point, which is also called Nilkamal, is a UNESCO World Heritage site.
11. Like Bangladesh, districts in India are local administrative units inherited from the British Raj. They are administrative division of an India state or territory.
12. Experts also believe that if the 2010 Haiti Earthquake had struck during the night or earlier in the day when more people would have been at school or work, the fatalities likely would have been even higher.

References

Ahsan, M.N., K. Takeuchi, K. Vink, and M. Ohara. 2016. A Systematic Review of the Factors Affecting the cyclone Evacuation Decision Process in Bangladesh. *Journal of Disaster Research* 11(4): 742–753.

AP (Associated Press). 2008. Cyclone Nargis Embodied the 'Perfect Storm.' 8 May (www.nbcnews.com/id/24526960/ns/world_news-asia_pacific/t/cyclone-nargis-embodied-perfect-storm/#.XUsTnfJKhhE – last accessed August 7, 2019).

BSL (Berkeley Seismology Lab). 2015. Today in Earthquake History: Chile 1960, 22 May (https://seismo.berkeley.edu/blog/2015/05/22/today-in-earthquake-history-chile-1960.html – last accessed August 12, 2019).

Blake, G. 2008. The Gathering Storm. *OnEarth* 30(2): 22–37.

Bolitho, S. 2015. Tropical Cyclone Pam: Why the Vanuatu Death Roll Was so Low. ABC News, 1 April.

Brumbaugh, D.S. 1999. *Earthquakes: Science and Society*. Upper Saddle River, NJ: Prentice Hall.

CRED (Center for Research on the Epidemiology of Disasters) and UNISDR (The United Nations Office for Disaster Risk Reduction). 2016. *Poverty & Death: Disaster Mortality 1996–2015*. Brussels: CRED.

Chowdhury, A.M.R., A.U. Bhuyia, A.Y. Choudhury, and R. Sen. 1993. The Bangladesh Cyclone of 1991: Why So Many People Died. *Disasters* 17(4): 291–304.

City-Data.com. 2011: Joplin, Missouri (www.city-data.com/city/JoplinMissouri.html – last accessed March 15, 2012).

Coles, J.B., J. Zhuang, and J. Yates. 2012. Case Study in Disaster Relief: A Descriptive Analysis of Agency Partnerships in the Aftermath of the January 12th, 2010 Haitian Earthquake. *Socio-Economic Planning Sciences* 46: 67–77.

CRS (Congressional Research Service). 2014. *Typhoon Haiyan (Yolanda): U.S. and International Response to Philippines Disaster*. Washington, DC: CRS.

Curtis, A., and J.W. Mills. 2009. *GIS, Human Geography, and Disasters*. San Diego: University Readers.

Daniell, J.E., B. Khazai, and F. Wenzel. 2013. Uncovering the 2010 Haiti Earthquake Death Toll. *Natural Hazards Earth System Science* 1: 1913–1942.

Ebert, C.H.V. 1993. *Disasters: Violence of Nature Threats by Man*. Dubuque, IA: Kendall/Hunt Publishing Company.

Fetter, J. 2017. Haiti vs. Chile: The Earthquake Olympics. Huffpost, 6 December (www.huffpost.com/entry/haiti-vs-chile-the-earthq_b_518639?guccounter=1&guce_referrer=aHR0cDovL3NlYXJjaC50Yi5hc2suY29tL3NlYXJjaC9HR21haW4uamh0bWw_c2VhcmNoZm9yPUNvbXBhcmlvbitvZisyMDEwK0hha – last accessed August 11, 2019).

Frank, N.L., and S.A. Husain. 1971. The Deadliest Tropical Cyclone in History. *Bulletin of the American Meteorological Society* 52(6):438–445.

Fuller, P. n.d. Haiyuan Earthquake, 1920 (www.disasterhistory.org/gansu-earthquake-1920 – last accessed November 17, 2018).

GoB (Government of Bangladesh). 2008. *Cyclone Sidr in Bangladesh: Damage, Loss and Needs Assessment for Disaster Recovery and Reconstruction*. Dhaka: GoB.

GoRV (Government of the Republic of Vanuatu). 2015. *Vanuatu Post-Disaster Needs Assessment: Tropical Cyclone Pam, March 2015*. Port Vila, Vanuatu: GoRV.

Handmer, J., and H. Iveson. 2017. Cyclone Pam in Vanuatu: Learning from the Low Death Toll. *Australian Journal of Emergency Management* 32(2): 60–65.

Haque, C.E. 1995. Climatic Hazards Warning Process in Bangladesh: Experience of, and Lessons from, the 1991 April Cyclone. *Environmental Management* 19(5): 719–734.

Haque, C.E., and D. Blair. 1992. Vulnerability to Tropical Cyclones: Evidence from the April 1991 Cyclone in Coastal Bangladesh. *Disasters* 16: 217–229.

Haque, U M. Hashizume, K.N. Kolivras, H.J. Overgaard, B. Das, and Taro. Yamamoto. 2012. Reduced Death Rates from Cyclones in Bangladesh: What More Needs to be Done? *Bulletin of the World Health Organization*, 2012 (90): 150–156.

Hossain, M.Z., M.T. Islam, T. Sakai, and M. Ishida. 2008. Impact of Tropical Cyclone on Rural Infrastructure in Bangladesh. *Agricultural Engineering International* 10(2): 1–13.

Ikeda, K. 1995. Gender Differences in Human Loss and Vulnerability in Natural Disasters: A Case Study from Bangladesh. *Indian Journal of Gender Studies* 2(2): 171–193.

IFRC (International Federation of Red Cross and Red Crescent Society). 2010. *Final Report: Bangladesh Cyclone Aila*. Dhaka: IFRC.

Joplin Globe. 2011. Following May 22, How Can Homeowners Protect Themselves for the Next One? June 26.

Karan, P.P. 2011. Introduction: When Nature Turns Savage. In *The Indian Ocean Tsunami: The Global Response to a Natural Disaster*, edited by Karan, P.P., and S.P. Subbiah, pp. 1–13. Lexington: The University Press of Kentucky.

Kirsch, T.D., E. Leidman, W. Weiss, and S. Dooey. 2012. The Impact of the Earthquake and Humanitarian Assistance on Household Economies and Livelihoods of Earthquake-Affected Populations in Haiti. *American Journal of Disaster Medicine* 7(2): 85–94.

Kte'pi, B. 2011. Haiyuan Earthquake (1920). In *Encyclopedia of Disaster Relief*, edited by Penuel, B.K., and M. Statler, pp. 261–263. Thousand Oaks, CA: SAGE.

McCullough, D.G. 1968. *The Johnstown Flood*. New York, NY: Simon and Schuster.

McPherson, M., M. M. Counahanb, and J.L. Hallb. 2015. Responding to Typhoon Haiyan in the Philippines. *Western Pacific Surveillance and Response Journal* 6(S 1): 1–4.

Mersereau, D. 2013. Why So Many People Died from Haiyan and Past Southeast Asia Typhoons, 11 November. *The Washington Post* (www.washingtonpost.com/news/capital-weather-gang/wp/2013/11/11/inside-the-taggering-death-toll-from-haiyan-and-other-southeast-asia-typhoons/ – last accessed August 6, 2019).

Montz, B.E., G.A. Tobin, and R.R. Hagelman, III. 2017. *Natural Hazards: Explanation and Integration*. New York, NY: The Guilford Press.

Mustain, A. 2011. 2011 Tornado Death Toll is Worst Since 1953. LiveScience (www.livescience.com/14294-2011-tornado-death-toll-worst-1953.html – last accessed February 11, 2012).

NWS (National Weather Service). 2011. NWS *Central Region Service Assessment: Joplin, Missouri, Tornado* (www.nws.noaa.gov/om/assessments/pdfs/Joplin_tornado.pdf – last accessed March 11, 2012).

NWS (National Weather Service). 2020. *Preliminary US Flood Fatality Statistics* (weather.gov/arx/usflood – last accessed May 20, 2020).

Ozerdem, A. 2006. The Mountain Tsunami: Afterthoughts on the Kashmir Earthquake. *Third World Quarterly* 27: 397–419.

Padhy, G., R.N. Padhy, S. Das, and A. Mishra. 2015. A Review on Management of Cyclone Phailin: Early Warning and Timely Action Saved Life. *Indian Journal of Forensic and Community Medicine* 2(1): 56–63.

Pallardy, R. n.d. 2010 Haiti Earthquake. *Encyclopedia Britannica* (www.britannica.com/event/2010-Haiti-earthquake – last accessed August 3, 2019).

Pallardy, R., and J.P. Rafferty. n.d. Chile Earthquake of 2010. *Encyclopedia Britannica* (www.britannica.com/event/Chile-earthquake-of-2010 – last accessed August 11, 2019).

Palmer, J. 2012. *Heaven Cracks, Earth Shakes: The Tangshan Earthquake and the Death of Mao's China*. New York, NY: Basic Books.

Paul, B.K. 2009. Why Relatively Fewer People Died? The Case of Bangladesh's Cyclone Sidr. *Natural Hazards* 50: 289–304.

Paul, B.K. 2011. *Environmental Hazards and Disasters: Contexts, Perspectives and Management*. Hoboken, NJ: Wiley-Blackwell.

Paul, B.K. 2013. Religious Interpretations for the Causes of the 2004 Indian Ocean Tsunami. *Asian Profile* 41(1): 67–77.

Paul, B.K. 2019. *Disaster Relief Aid: Changes & Challenges*. Gewerbestrasse, Switzerland: Palgrave Macmillan.

Paul, B.K., and S. Dutt. 2010. Hazard Warnings and Responses to Evacuation Orders: The Case of Bangladesh's Cyclone Sidr. *Geographical Review* 100(3): 336–355.

Paul, B.K., and Stimers, M. 2012. Exploring Probable Reasons for Record Fatalities: The Case of 2011 Joplin, Missouri, Tornado. *Natural Hazards* 64(2): 1511–1526.

Paul, B.K., and S. Chatterjee. 2019. Climate Change-Induced Environmental Hazards and *Alia* Relief Measures Undertaken to Sundarbans in Bangladesh and India. In *The Sundarbans: A Disaster-Prone Eco-Region, Increasing Livelihood Security*, edited by H.S. Sen, pp. 471–490. Gewerbestrasse, Switzerland: Springer.

Paul, B.K., and Ramekar, A. Host Characteristics as Risk Factors Associated with the 2015 Earthquake-Induced Injuries in Nepal: A Cross-Sectional Study. *International Journal of Disaster Risk Reduction* 27 (2018): 118–126.

Paul, B.K., H. Rashid, M.S. Islam, and L.M. Hunt. 2010. Cyclone Evacuation in Bangladesh: Tropical Cyclones Gorky (1991) vs. Sidr (2007). *Environmental Hazards* 9: 89–101.

Paul, B.K., and M. Stimers. 2014. Spatial Analyses of the 2011 Joplin Tornado Mortality: Deaths by Interpolated Damage Zones and Location of Victims. *Weather, Climate and Society* 6(2): 161–174.

Paul, B.K., M. Stimers, and M. Caldas. 2015. Predictors of Compliance with Tornado Warnings Issued in Joplin, Missouri, in 2011. *Disasters* 39(1): 108–124.

Penna A.N., and J.S. Rivers. 2013. Natural Disasters in a Global Environment (https://ebookcentral-proquest-com.ezproxy.lib.vt.edu – last accessed March 19, 2018).

Rashid, H., and B.K. Paul. 2014. *Climate Change in Bangladesh*. New York, NY: Lexington Books.

Ray-Bennett, N.S. 2016. Learning from Deaths in Disasters: The Case of Odisha, India, June 7 (www.mei.edu/publications/learning-deaths-disasters-case-odisha-india – last accessed August 5, 2019).

Reuters. 2009. FACTBOX-Key facts about Cyclone Nargis, 30 April (www.reuters.com/article/idUSSP420097 – last accessed August 7, 2019).

Robinson, S. 2007. How Bangladesh Survived the Cyclone. *Time Magazine*, 19 November.

Ryan, K. 2011: Joplin School District Looks for Way Forward. *Joplin Globe*, 24 May (www.joplinglobe.com/local/x1439575311/Joplin-School-District-looks-for-wayforward – last accessed March 14, 2012).

Salley, J. 2017. Remembering the 1970 Bhola Cyclone, the World's Deadliest Weather Event, May 18 (www.necn.com/news/national-international/Remembering-the-1970-Bhola-Cyclone-422996194.html – last accessed August 1, 2019).

Samenow, J. 2013. Major Disaster Averted: 5 Reasons Why Cyclone Phailin Not As Bad As Feared in India. *The Washington Post*. October 14.

Shamsuddoha, M., and R.K. Chowdhury. 2007. *Climate Change Impact and Disaster Vulnerabilities in the Coastal Areas of Bangladesh*. Dhaka: COAST Trust.

Shelter Cluster. 2015. *Tropical Cyclone Pam Response* (www.sheltercluster.org/sites/default/files/docs/vanuatu_sc_ll_final_report_v2_22062015.pdf. – last accessed August 14, 2019.

Simmons, K.M., and D. Sutter. 2011. *Economic and Societal Impacts of Tornadoes*. Boston: American Meteorological Society.

Simmons, K.M., and D. Sutter. 2012. *Deadly Season: Analysis of the 2011 Tornado Outbreaks*. Boston: American Meteorological Society.

Singh, D., and A. Jeffries. 2013. *Cyclone Phailin in Odisha, October 2013: Rapid Damage and Needs Assessment Report*. Washington, DC: World Bank.

Stimers, M. J. 2011: A Categorization Scheme for Understanding Tornado Events from the Human Perspective. Ph.D. dissertation, Department of Geography, Kansas State University.

Stimers, M.J., and B.K. Paul. 2017. Deaths as a Function of Elevation: The Joplin, MO, Tornado, May, 2011. *Journal of Geography and Natural Disasters* 7(3): 1–7.

SPC (Storm Prediction Center). 2012: Annual U.S. Killer Tornado Statistics (www.spc.noaa.gov/climo/torn/fataltorn.html – last accessed March 23, 2012).

U.S. Census Bureau. 2012. *Missouri 2010: Population and Housing Unit Counts*. Washington, DC: US Department of Commerce.

USGS (United States Geological Survey) and American Red Cross. 2011. *Report on the 2010 Chilean Earthquake and Tsunami Response*. Washington, DC: U.S. Department of Interior and U.S. Geological Survey.

Vijaykumar, D. 2015. What Chile Did Right. *Reliefweb*, 18 September (https://reliefweb.int/report/chile/what-chile-did-right – last accessed June 12, 2020).

Wayman, E. 2010. Chile's Quake Larger but Less Destructive than Haiti's, 1 March (www.earthmagazine.org/article/chiles-quake-larger-less-destructive-haitis – last accessed August 11, 2019).

Wurman, J., C. Alexander, P. Robinson, and Y. Richardson. 2007. Low-level Winds in Tornadoes and Potential Catastrophic Tornado Impacts in Urban Areas. *Bulletin of American Meteorological Society* 88. 31–46.

Yamada, S., R. Gunatilake, T.M. Roytman, S. Gunatilake, and T. Fernando. 2006. The Sri Lanka Tsunami Experience. *Disaster Management & Response* 4(2): 38–48.

Yamada, S. 2017. Hearts and Minds: Typhoon Yolanda/Haiyan and the Use of Humanitarian Assistance/Disaster Relief to Further Strategic Ends. *Social Medicine* 11(2): 76–82.

3 Trends and levels of disaster deaths

This chapter addresses the first objective of this book, which is to explore the trends and levels of disaster-induced deaths to find out whether such deaths have decreased or increased in the period considered across all disaster types and geographical scales. Where different disasters show different trends at different scales, the chapter attempts to explain such trends. Analysis of disaster trends is necessary because the number of deaths from extreme natural events is highly variable from year to year; some years have very few deaths, while other years deadly disasters claim many lives. This type of analysis will provide useful insights about which disasters and geographic scales need stronger interventions to reduce disaster-induced deaths. Because many countries have intensified their efforts to reduce disaster deaths by implementing many preparedness and mitigation measures in the last three decades or so, we assume a downward trend in disaster mortality across the scales and types of extreme events. If the assumption proves to be incorrect, reasons for that resulting trend are explored, and recommendations are made to reverse the trend. A quantitative approach is used to analyze the trends of disaster mortality.

Trends and levels of disaster deaths are analyzed in this chapter for the 25-year period (1991–2015). The period was selected primarily to maintain consistency in the source of information. The data source is the three World Disasters Reports (IFRC 2001, 2011, 2016) published by the International Federation of Red Cross and Red Crescent Societies (IFRC).[1] These reports used the Center for Research on the Epidemiology of Disasters (CRED) Emergency Events Database (EM-DAT) for all information related to natural and technological disasters. Despite some limitation, EM-DAT is the most comprehensive global database on natural and technological disasters.[2] Moreover, the IFRC has been publishing World Disasters Reports (WDR) annually since 1993, and each WDR publishes disaster data for the previous 10 years in tabular format in the annex. Recently, the WDR discontinued presenting relevant data in tabular form in favor of providing a graphical summary of the data with reference to the previous decade. However, earlier reports are not readily available, and therefore the analysis of disaster deaths is presented for the 1991–2015 period.

Levels of deaths here simply refer to the number of direct and indirect deaths caused by a natural disaster in the area affected. No attempt is made to calculate the crude death rate (CDR) (meaning the total deaths divided by the number of people in the disaster-affected area, and then multiplied by 100, 1,000, or another appropriate number). The major problem with using the death rate is the non-availability of population data for the disaster-affected areas. Country-wide population data are readily available, but a natural disaster hardly ever afflicts the entire nation. For the same reason, age-specific death rates, which are more meaningful than CDR, are not used.

However, both trends and levels of disaster deaths are analyzed according to the following: (1) all natural disasters together; (2) nine individual disaster types (avalanches/landslides, droughts/famines, earthquakes, extreme temperatures, floods, forest/wild fires, volcanic eruptions, wind storms (e.g., tropical cyclones, tornadoes, thunderstorms, dust storms, hail storms, and severe storms), and all other natural disasters); (3) two major types (climate-hydro-meteorological and geophysical disasters); (4) those on five continents (Africa, Americas, Asia, Europe, and Oceania); and (5) those based on three world regions (high human development, medium human development, and low human development) as identified each year by the United Nations Development Program (UNDP). For convenience, the selection of these sub-groups is based on EM-DAT format provided in the WDRs. All analyses are performed for the entire study period as well as only for the last five years (2011–2015) of the study period to examine the recent status of mortality in comparison to a relatively long period.

Note that the focus of this book is exclusively on natural disasters. It excludes biological and technological disasters, complex emergencies, and terrorisms. Even within the natural disasters, minor hazards such as fog, frost, sinkholes, subsidence, water spouts, and (coastal) erosion are not considered because of their contribution in causing large number of deaths. Meteorites or extinction are not included because they are considered rare events.

However, the latter part of this chapter also covers two interrelated topics of disaster-induced deaths. The first topic addresses one of the disaster death myths that dead bodies cause epidemics immediately after extreme events occur (de Goyet 2000; Morgan et al. 2006). This fear is the principal cause of rapid burial of human remains in mass graves frequently without identification of the dead. This action has psychological implications for surviving family of deceased persons, and includes other implications such as social, economic, and legal problems affecting inheritance, compensation, insurance, and re-marriage of spouses (Perera 2005). Another related topic this chapter covers is the stages and issues linked to mass-fatality caused by large natural disasters and the management of such fatality.

Disaster death trends

Two simple statistics are used to examine the number of deaths caused by natural disasters during the study period (1991–2015). This is because deaths from natural disasters can be highly variable from one year to another. To measure this variability, the coefficient of variation (CV), the ratio of the standard deviation to the mean, is used here. The CV is a relative measure of dispersion and is usually expressed either as a proportion or percentage. It allows direct comparison of relative variability in different data sets (McGrew Jr. et al. 2014). The higher the CV value, the greater the level of dispersion around the mean. In other words, a lower CV implies a low degree of variation while a higher CV connotes a higher variation. Distributions with a CV less than 1 are considered to be low variance, whereas those with a CV higher than 1 are considered to be high variance.

Secondly, to examine the trend of deaths caused by natural disasters, a series of bivariate regressions is used where the independent variable (x) is the year, and the dependent variable (y) is the number of deaths. Regression coefficient or slope (b-value) will indicate whether the trend is positive or negative over the study period. If the b-value is negative, it represents a negative trend, and if it is positive, the trend is positive; i.e., the number of deaths is increasing over the study period.

Aggregate disaster deaths trends

In the study period (1991–2015), all natural disasters claimed 1.99 million lives worldwide or 79,617 per year[3] (Figure 3.1). This is slightly higher than the annual 68,000 deaths reported by CRED, USAID, and UNISDR (2016) in their recent report for the period 1994–2013. This higher number of annual deaths is associated with several major deadly disasters that occurred before

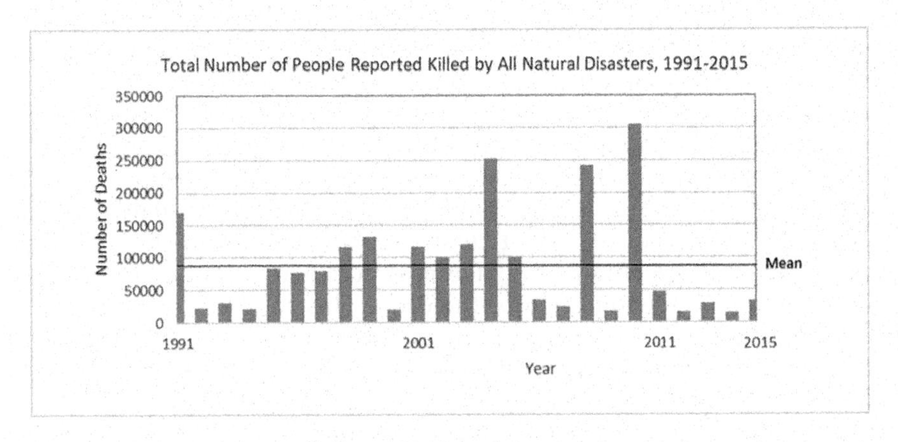

Figure 3.1 Worldwide aggregate disaster deaths by year, 1991–2015.

Source: IFRC (2001, 2011, and 2016).

or after the period considered by the CRED, USAID, and UNISDR (2016). For example, Cyclone Gorky made landfall in coastal Bangladesh in 1991 and killed nearly 140,000 people (Paul et al. 2010). Also, Nepal earthquake occurred in 2015 and killed at least 8,857 people (GoN 2015; Paul et al. 2017). However, when the last five years (2011–2015) of the present study period is considered, annual deaths due to all natural disasters come to 20,700 fatalities, which is much lower than reported by CRED, USAID, and UNISDR (2016) for the 1994–2013 period. When the most recent years (2016–2019) are considered, the average annual number of deaths even reduced to 9,459 (Ritchie and Roser 2019; Statista 2020). Moreover, Ritchie and Roser (2019) analyzed the global deaths from all natural disasters for 1978–2018, and found a rapidly declining trend for the period.[4]

The greatest number of deaths during the present study period occurred in 2010 when 297,752 people died from all natural disasters, and the lowest number of deaths occurred in 2014 when 8,000 people died. In particular, 2010 was the deadliest year in more than a generation. More people were killed worldwide by natural disasters in 2010 because several major disasters of different types occurred: earthquakes (e.g., Haiti), heat waves (e.g., Russia), floods (e.g., Pakistan, India, Italy, Colombia, Chad, and the USA), typhoons (e.g., Typhoon Conson in the Philippines), blizzards (e.g., USA), volcanic eruptions (e.g., Iceland), landslides (e.g., Haiti and Uganda), and droughts (e.g., USA). Of these events, the Haitian earthquake, Russian heat wave, and Pakistani flooding were the biggest killers, accounting for 96 percent of all deaths caused by natural disasters in 2010 (IFRC 2011). However, deadly earthquakes also struck Chile, Turkey, China, and Indonesia, while 26 people died in the Tennessee floods in the United States. Although floods (e.g., India, Pakistan, Solomon Islands, and Bosnia-Herzegovina), typhoons (e.g., the Philippines), and earthquakes (e.g., China) occurred in 2014, no natural disasters accounted for a very large number of deaths in 2014.

Figure 3.1 shows the disaster deaths in each year of the study period. The figure illustrates that the number of deaths was relatively the highest in the first decade of the twenty-first century (2001–2010) where the number of deaths was above average in seven of the 10 years. This is not unexpected since a number of major disasters occurred during this decade. Notable among these were the 2001 Gujarat, India, earthquake, the 2003 Bam, Iran, earthquake, and the Summer heatwave in Europe, the 2004 Indian Ocean Tsunami (IOT), the 2005 Kashmir, Pakistan, earthquake, the 2008 Cyclone Nargis in Myanmar, and the 2010 Haiti earthquake and Pakistani floods. The second highest deaths were recorded in the 1990s, where four years experienced above annual average deaths of 79,617. Several major natural disasters occurred during this decade such as the 1990 Northwestern Iran earthquake, the 1991 Cyclone Gorky in Bangladesh, the 1995 Kobe earthquake in Japan, and the 1999 Western Turkey earthquake.

However, the number of deaths remained below the yearly average in each year of the last five years (2011–2015) of the study period. Figure 3.1 further shows that year-to-year fluctuations were not high during the entire study period; correspondingly, the CV is 0.993. A careful review of Figure 3.1 does not reveal any clear trend. However, the fitting of a linear trend line is negative ($b = -309.693$), indicating a negative trend over the study period. This means that the number of deaths caused by natural disasters was declining. This trend is consistent with the trend reported by Coppola (2007), who analyzed total number of deaths caused by all natural disasters worldwide for the 1900–2004 period (also see Ritchie and Roser 2019). He provided several reasons for this downward trend.

> Greater recognition of the importance of emergency management and sustainable development are turning this tide on disasters. The efforts of the UN, the many nongovernmental agencies involved in development and disaster preparedness and response, and the efforts of individual governments have shown that humans can effectively influence their vulnerability.
>
> *(Coppola 2007, 17)*

Individual disaster deaths trends

In the WDR, the IFRC considers nine different types of natural disasters. The number of deaths in each disaster type is presented in descending order with regard to contribution of each type to the total deaths caused by all nine disasters during the study period (1991–2015). Earthquakes (including tsunamis) were the deadliest type of disaster, accounting for nearly 39 percent of total disaster deaths (Table 3.1). This translates to 772,470 deaths during the study period. Two mega events (the 2004 IOT and the 2010 Haiti earthquake) during the study period accounted for 454,025 (or 59 percent) of these 772,470 deaths. The overwhelming majority of the deaths in 2004 were

Table 3.1 Worldwide deaths caused by all natural disasters during 1991–2015 by individual event

	Deaths			*Death trend*	
Disaster	*Percentage*	*Annual average*	*CV value*	*b-value*	*Direction*
Earthquakes	38.81	35,709	2.0	1,771.81	Increased
Droughts	22.30	29,590	1.27	−2470.25	Decreased
Wind storms	19.87	15,819	2.45	−886.30	Decreased
Floods	9.02	7,223	0.82	−154.01	Decreased
Extreme Temp.	8.51	6,773	2.62	395.31	Increased
Avalanches	1.14	598	1.39	−5.07	Decreased
Volcanic Erupt.	0.08	64	2.29	−5.46	Decreased
Forest Fires	0.08	60	0.81	−0.06	Decreased
Others	0.15	123	3.49	−9.96	Decreased

Sources: IFRC (2001, 2011, and 2016).

caused by the tsunami. No other tsunami killed that many people during the study period.[5] However, on average, 30,899 people annually died from earthquakes and tsunamis during the study period. When the last 10 years of the study period is considered, 35,709 people died annually from these events. The CV value of 2.0 shows that earthquake and tsunami deaths fluctuated during the study period primarily for the unexpectedly large deaths that occurred in 2004 and 2010.

The *b*-value shows that the number of deaths caused by earthquakes and tsunamis increased over the study period (Table 3.1). This increase is largely explained by the absence of an effective early earthquake warning system and by deadly earthquake events occurring in 2010 in Haiti and in 2015 in Nepal. Another reason for the high number of deaths is the expansion of urbanization and rapid population growth within highly seismic zones in the last few years. Slums and squatter settlements frequently expand on the highest risk areas such as areas with steep slopes (CRED, USAID, and UNISDR 2016). It must be noted that ground shaking during an earthquake is seldom the direct cause of death or injury. Most earthquake-related fatalities result from building collapse or damage. This is largely because buildings in most of the urban areas, particularly in developing countries, are not able to resist the force generated by seismic waves because of low-quality building construction.

Droughts (with famines and associated food insecurities) are the second deadliest natural disaster.[6] EM-DAT shows that 443,844 people died from droughts during the study period, resulting in an average 29,590 fatalities per year. For the last five years of the study period, the number of deaths caused by droughts was only 2,007 per year. Conversely, the greatest number of deaths occurred in 2002 when 76,903 people died from drought events. In six years (1993, 1994, 2007, 2012, 2013, and 2014), no deaths from droughts were recorded, whereas the number of deaths above the yearly average was reported in seven years. Five of these seven years were in the decade of the 1990s. This decade also experienced a higher number of deaths than the subsequent decade.

Droughts not only cause water and food shortages but also have many health impacts, which may increase morbidity that results in death. Droughts contributed to 22.30 percent of the deaths caused by all natural disasters during the study period (1991–2015). However, the calculated CV presented in Table 3.1 shows that annual drought fatalities have been somewhat variable; nevertheless, the trend line ($b = -2470.25$) clearly shows a negative trend during the entire study period (Table 3.1). Unfortunately, this trend may be reversed due to potential impacts of climate change if the countries are unwilling to combat climate change.

Storms (including cyclones/hurricanes/typhoons, tornadoes, blizzards, thunderstorms and lightning, dust storms, and tropical storms) were the third deadliest type of disaster accounting for nearly 20 percent of disaster deaths (Table 3.1). Among all extreme events included under storms, the number one killer was tropical cyclones. However, all storm-related disasters killed 395,476 people during the period studied. Nearly 73 percent of the 395,476 deaths were caused by two mega-disasters: 1991 Cyclone Gorky in Bangladesh and 2008 Cyclone Nargis in Myanmar. The third mega disaster, the 2010 Haiti earthquake, killed close to 230,000 people (Paul 2019). Note that the CRED defines a mega-disaster as an event that kills more than 100,000 people (CRED, USAID, and UNISDR 2016).

Notably, with the exception of 1991 and 2008, on average, 4,675 people died annually; however, when these two years are included, this average jumped to 15,819 deaths per year. The CV value (2.45) is also highly influenced by the unexpectedly huge number of deaths that occurred in 1991 and 2008. Finally, the regression slope ($b = -886.30$) clearly indicated a downward trend in the number of deaths caused by windstorms during the study period (Table 3.1).

Worldwide, all types of floods (e.g., river, coastal or tidal, and flash floods) were responsible for slightly over nine percent of all deaths resulting from natural disasters from 1991 to 2015. This means that of the nearly 2 million deaths, 180,576 were caused by floods. This, in turn, means that during the study period, floods killed 7,223 people annually. The CV value (0.816) indicates that flood fatalities did not fluctuate much during the period, but that they ranged from 3,408 in 2015 to 34,366 in 1999 (Table 3.1). The highest number of deaths occurred in 1999 mainly due to two floods: the Venezuela and Columbia floods together killed 20,000 people, and 9,863 people died due to floods in India. Additionally, 98 other major floods occurred in 1999 in 45 other countries, including the United States. These latter floods killed at least 2 persons, while eight of them killed more than 100 people.

Floods were less deadly than earthquakes, droughts, or windstorms in terms of the numbers of lives lost during the study period; moreover, the *b*-value (−154.01) clearly suggests that floods had been becoming less deadly during the study period (Table 3.1). This does not necessarily mean that all flood-prone countries of the world showed a declining trend. However, the declining trend might be associated with installation and improvement of early flood warning systems, along with provision to evacuate people from flood-risk zones. It is also partially explained by regular improvement of river channels and construction of maintenance of flood defenses.

Extreme temperatures, including heat waves and cold waves, killed 169,332 people or 6,773 per year. The number of deaths was highest (74,748) in 2003 followed by 2010 (57,268). First, the mega heatwave of Europe in 2003 killed people ranging between 35,000 and 70,000 (Stott et al. 2004). France was hit especially hard where at least nearly 15,000 people died. Other European countries affected by this heat wave were Germany, Ireland, Italy, the Netherlands, Portugal, Spain, Sweden, Switzerland, and the United Kingdom. Notably, 2010 was the fourth warmest summer in the 131-year temperature record both in parts of Europe (e.g., Eastern Europe and Russia) and the United States.[7] In contrast, the number of deaths was lowest in 1993 when only 106 people died worldwide.

In some countries, such as the United States, more deaths are attributed to extreme temperatures than to flood, hurricanes, tornadoes, earthquakes, and lightning. Overall, extreme temperatures accounted for 8.5 percent of all deaths caused by natural disasters during the study period. However, the CV value (2.619) suggests that the number of deaths varied greatly because of the unusually large numbers in 2003 and 2010. Overall, number of deaths increased during the study period (Table 3.1) primarily because of the increase in heat-related deaths, which are likely to increase significantly as global temperatures rise due to increased greenhouse gas emissions (GGE). Scientists believe that a large increase in temperature-related deaths can be reduced by decreasing GGE and by putting in place mitigation and adaptation measures.

During the 25-year study period, avalanches/landslides and mass movement (wet) killed 22,562 people; this represents only 1.13 percent of all deaths caused by natural disasters in the study period. Unlike droughts, avalanches killed every year during the study period, and the lowest number of people died (235) in 2013, far below the annual average of 598 people (Table 3.1). The highest number of deaths occurred in 2010 when 3,402 people died worldwide. Thus, avalanches are not the largest contributor to deaths, but there have been a few events in history that have killed thousands of people. For example, the deadliest avalanches occurred in May/June in Peru, in 1970 and killed 20,000 people. It caused collapse a substantial section of the north slope of Mt. Huascaran. Overall, the number of deaths caused by avalanches/landslides has been declining over time (Table 3.1).

Volcanic eruptions contributed to 0.08 percent of all deaths caused by natural disasters during the study period. In 10 of the 25 study years (1995, 1998–2001, 2003, 2009, 2012, 2013, and 2015), no mortality was reported for volcanic eruptions. The number of deaths range from 2 in 1992 to 683 in 1991.[8] In fact, 55 volcanic eruptions caused the deaths of 460 people from 2006 to 2015 (CRED, USAID, and UNISDR 2016). Unlike many natural disasters, these events provide advance warning signs of imminent eruptions, which partially explain relatively quite low volcanic-induced fatalities. Another reason for low mortality is the remote location of most of the volcanoes in the world.

Although deaths induced by forest or wildfires occurred in every year of the study period, annual average deaths caused by forest fires was nearly 60 persons as opposed to slightly over 64 deaths per year for volcanic eruptions (Table 3.1). There is also another difference between these two types of disasters. The annual variation in the number of deaths was higher in the cases of volcanic eruptions than for forest/wildfires. Despite higher frequencies of wildfires in recent years, particularly in the United States and Australia, the number of deaths caused by both types of disasters shows a declining trend over the study period (Table 3.1).

Other minor disasters such as insects, waves/surges, and dry-mass movements, are considered as "others," and accounted for 3,064 deaths during the study period. These minor disasters annually killed nearly 123 people per year. But this "others" category did not result in the death of anyone in nine of the 25 years, and in 13 years, fewer than 100 people died. Thus, the year to year variability is very high (CV = 3.49), but the number of deaths during the study period shows a downward trend (Table 3.1).

Deaths trends by two broad types of natural disasters

EM-DAT broadly classified nine individual natural disasters into two groups: climate-related and geophysical/geological disasters. The former includes hydrological, meteorological, and climatological disasters, while the latter includes events such as earthquakes, tsunamis, landslides (dry), coastal erosions, avalanches, lahars, and volcanic eruptions. Tsunamis are considered under geophysical events because their root cause is seismic activity (CRED, USAID, and UNISDR 2016). In fact, geophysical events are disasters that originate by tectonic and seismic activities below the earth's surface. Of the total deaths during the study period, climate-related disasters accounted for 61 percent of the total deaths, and the remaining 39 percent of deaths were caused by geophysical events (Table 3.2).[9] This is not a surprising finding because climate-related events accounted for the overwhelming majority (91 percent) of natural disasters between 1991 and 2015.[10] This means that geophysical disasters accounted for a much larger share of the total deaths relative to the share of their number of reported disasters. Thus, geophysical disasters, particularly earthquakes, are low-frequency, but high impact events.

Table 3.2 clearly shows that year-to-year variation in the number of deaths is much less for the climate-related disasters than for the geophysical disasters, which is not unusual for high-frequency and low impact events. This means some years recorded relatively low numbers of deaths before a large disaster event claimed many lives. This is also reflected in CV values; the CV value for climate-related events during the study period is 0.92, while for the geophysical events, the corresponding CV is 2.00. Similarly, different trends are observable between these two broad types of disasters: climate-related disasters show a negative trend, while geophysical events show a positive trend (Table 3.2). This trend probably resulted from the earthquakes that

Table 3.2 Worldwide deaths caused by two broad types of disasters during 1991–2015

	Deaths			*Death trend*	
Broad disaster type	*Percentage*	*Annual average*	*CV value*	*b-value*	*Direction*
Climate relate	61.09	48,640	0.92	−208,572	Decreased
Geological	38.91	30,976	2.00	1,717	Increased

Sources: IFRC (2001, 2011, and 2016).

occurred in 2010, killing nearly 321,000. Most of these deaths were due to the Haiti earthquake. In the following year, the Tohoku earthquake and tsunami in Japan killed between 18,000 and 20,000 people. Then, the 2015 earthquake in Nepal killed more than 8,857 people (GoN 2015; Paul et al. 2017).

Disasters death trends by continent

Among the five continents considered (Africa, Americas, Asia, Europe, and Oceania), Asia is the most disaster-prone. The continent accounted for 70.38 percent of all deaths caused by natural disasters during the study period, followed by the Americas (15.68 percent), Europe (8.73 percent), Africa (4.92 percent), and Oceania (0.28 percent) (Table 3.3). When considered with respect to its share of total disasters during the study period and its share of total population of the world, Asia's number of deaths is higher proportionally for both metrics. During the study period, nearly 42 percent of natural disasters occurred in Asia, and the continent accounts for nearly 60 percent of the world population. This means relatively more deadly disasters occurred in Asia: 1991 Cyclone Gorky in Bangladesh; 1995 Kobe earthquake in Japan; 2001 Gujarat earthquake in India; 2003 Bam earthquake in Iran; 2004 IOT in Southeast and South Asia; 2005 earthquake in Pakistan; 2008 Cyclone Nargis in Myanmar; 2010 floods in Pakistan; 2011 Triple disasters in Japan; and 2015 earthquakes in Nepal.

The frequency of large and deadly disasters is not uncommon in Asia because of its large and varied landmasses. It has not only mountains, but its east coast lies along very active seismic and volcanic zones, known as "Ring of Fire."[11] Additionally, the Pacific Ocean and the Indian Ocean are the world's major breeding grounds for tropical cyclone/typhoons. Meanwhile, its multiple large river basins (e.g., the Ganges-Brahmaputra-Meghna, the Indus, the Irrawaddy, the Mekong, the Yangtze, and the Yellow River basins) are flood-prone. Additionally, widespread poverty and high population densities make it vulnerable to natural disasters. Although all types of natural disasters occur there, most people die from earthquakes as many

Table 3.3 Deaths caused by all natural disasters during 1991–2015 by continent

	Death			*Death trend*	
Continent	*Percentage*	*Annual average*	*CV value*	*b-value*	*Direction*
Africa	4.92	4,343	0.55	43.12	Increased
Americas	15.68	13,830	3.18	1009.89	Increased
Asia	70.38	62,069	1.04	−1715.48	Decreased
Europe	8.73	7,703	2.27	272.36	Increased
Oceania	0.28	245	1.82	−3.93	Decreased

Sources: IFRC (2001, 2011, and 2016).

Asian countries are highly vulnerable such as China, India, Indonesia, Iran, Nepal, Pakistan, and Turkey. Also, death per earthquake event was highest (1,373 people in 1980–2008) on the continent followed by storms (412 persons per event) (PreventionWeb 2020).

Like Asia, both North and South America contributed relatively more deaths compared to their share of the world population. These two continents together accounted for slightly over 13 percent of the world population during the study period, but their share of total deaths caused by natural disasters was approximately 16 percent. In terms of countries, the United States reported the highest numbers of natural disasters as well as the highest numbers of deaths. This can be attributed to its large and heterogeneous landmasses. Borden and Cutter (2008) examined the deaths caused by natural disasters in the United States for the time period 1970–2004. Surprisingly, they found that heat or drought was the deadliest disaster followed closely by severe summer weather such as thunderstorms, wind, and hail. Clearly, along with China and India, the United States is among the countries of the world most affected by natural disasters.

Europe's share of disaster-induced deaths was almost 1 percent less than its population share (approximately nine percent versus 10 percent). Unlike the Americas and Asia, this continent lacks diversity of natural disasters. The most common disasters in Europe have been floods, strong storms, and heat waves, while countries such as Iceland, Italy, and Greece have been prone to volcanoes and earthquakes. But these three countries rarely get tsunamis. Hurricanes are not common in Europe, primarily because of its northerly location, and the adjacent oceans or seas are much less likely to see a tropical storm develop. Thus, Europe has fewer natural disasters. In fact, a number of European countries (e.g., Finland, Iceland, Malta, Norway, Poland, and Sweden) rank very low in disaster risk according to a profile developed by the United Nations University (IFRC 2001, 2011, and 2016). However, Table 3.2 does show an increasing trend in deaths caused by natural disasters in Europe, primarily because of the increasing trend in frequency of disasters in the recent past.

Africa accounted for approximately 17 percent of the world population during the study period, but its share of total deaths due to natural disaster was about 5 percent. This low contribution can be explained in several ways. First, the number of disasters was relatively less frequent, and additionally, the continent lacks diversity of extreme events. Floods and droughts are the most prevalent and impactful type of natural disasters, but these disasters are confined to few countries (e.g., Eritrea, Ethiopia, and Sudan),[12] and they do not kill a lot of people. For example, most floods in African countries have killed fewer than 50 in the past three decades. Only drought is widespread across the continent, particularly in southern Africa, the Horn of Africa, and the Sahel. It kills directly or indirectly a relatively large number of people. Rather, the main killers in Africa are civil conflicts, and epidemic and endemic diseases. Among the countries of Africa, Kenya and

Mozambique are likely the most vulnerable to natural disasters. Floods and tropical cyclones are common disasters in Mozambique.

Information presented in Table 3.3 shows that deaths caused by natural disasters in Africa have increased during the study period. This was primarily due to rapid increase of population and urbanization, and widespread poverty on the continent, which means that the number of people exposed to natural hazards and disasters continuously increased in Africa over time. Additionally, most governments of African countries do not have the resources to reduce disaster risk through preventive and mitigation measures.

As Oceania covers a very large water and land area, it experiences one of the highest diversities of natural disasters. Among them, some of the more frequent disasters are storms (including hurricanes), floods, earthquakes, droughts, volcanic eruptions, landslides, extreme temperatures, fires, tornadoes, tsunamis, and climate change-induced sea level rise (SLR). This region accounts for 0.55 percent of the world population but contributed 0.28 percent of all deaths caused by natural disasters during the study period. This is because it experiences only 13 disasters annually, on average, which is considered a very low frequency of occurrence. The most disaster-prone country of this region is Australia followed by Papua New Guinea, and New Zealand. While earthquakes and volcanic eruptions do not pose any threat to Australia, these two are very common in New Zealand. However, Table 3.3 suggests a declining trend in disaster-induced deaths during the study period.

Disaster death trends by Human Development Index (HDI)

The Human Development Index (HDI) was created by the UNDP in 1990. The HDI is a compound index, consisting of three variables: life expectancy at birth, mean of years of schooling, and gross nation income (GNI) per capita. These variables represent long and healthy life, knowledge, and a decent standard of living dimensions, respectively. The HDI yields a value between 0 (the lowest) and 1 (the highest) and is used to rank countries into four tiers of human development: very high HDI countries (0.800–1.00), high HDI countries (0.700–0.799), medium HDI countries (0.550–0.699), and low HDI countries (below 0.550). Previously, the UNDP used three tiers of classification for countries (high, medium, and low) prior to 2001. Here, to maintain consistency, three tiers are used, as very high and high HDI are merged into one category. Note that HDI was not calculated for all countries of the world because of non-availability of relevant data. Also, the number of countries in each tier differs year to year. For example, 105, 39, and 44 countries were in first (high), second (medium), and third (low) tiers, respectively in 2014 (UNDP 2015). But 106, 41, and 41 were in these three tiers in 2015 (UNDP 2016).

Table 3.4 presents information related to selected aspects of disaster deaths for the study period (1991–2015) by three tiers of HDI. Data show

Table 3.4 Deaths caused by all natural disasters during 1991–2015 by HDI tier

	Death			*Death trend*	
HDI tier	*Percentage*	*Annual average*	*CV value*	*b-value*	*Direction*
High	13.36	11,840	1.918	691.06	Increased
Medium	37.16	32,939	1.466	−218.31	Decreased
Low	48.98	43,413	1.294	−857.78	Decreased

Sources: IFRC (2001, 2011, and 2016).

both percentages of deaths and annual average number of deaths consistently increased with descending order of HDIs. This is not surprising since the high development index countries are mostly in Europe, where disaster deaths were lowest among the continents. Also, there is no strong relationship between the three tiers of HDI and a country's level of development. However, all CV values are higher than one (Table 3.4), meaning that the yearly deaths rate (or index) was highly variable. But with the exception of high HDI countries, the remaining two categories of countries showed declining trends over the study period.

The examination of temporal trends clearly shows that among nine broad types of natural disasters considered in this chapter, the number of deaths caused by only two (earthquake and extreme temperatures) increased during the study period. How to reduce deaths from these two and other extreme events are discussed in subsequent chapters. Among the continents, Africa, Americas, and Europe recorded increasing trends of disaster deaths. Available data suggests that America's increased death trend was possibly caused by the number of reported deaths resulted from earthquakes. Similarly, Europe's increased trend has resulted from increased trend of heatwave deaths, and Africa's increased trend was due to deaths caused by droughts. Countries of these three continents should invest efforts to reduce their share of deaths. Increasing trend of deaths for countries of high HDI is primarily influenced by trends of Americas and Europe.

Disaster deaths and outbreak of epidemics

There are at least 10 popular myths associated with major natural disasters (Jacob et al. 2008). One such myth is the outbreak of epidemics after disasters, particularly earthquakes, hurricanes, tidal waves, and floods. It is a common belief that these epidemics result from contamination by dead bodies from disasters, which, in turn, causes additional deaths. In the case of floods, contamination of existing water supplies is also the cause of outbreaks of epidemic diseases.[13] This belief is wrongly popularized by mass media, as well as some medical and disaster professionals (Morgan 2006). After any major disaster, media reports almost always stress the outbreak of epidemics. Thus, it is widely prevalent among many lay people, including some first responders, and workers and officials of government and nongovernmental organizations (NGOs). The Pan American Health Organization (PAHO) has been trying to eradicate this myth.

Two points need to be emphasized. First, the likelihood of massive post-disaster epidemics is generally grossly exaggerated. Although disasters damage or destroy the health care infrastructure, they rarely cause epidemics. In the case of geological disasters such as earthquakes, tsunamis, and landslides, the likelihood of emergence of an infectious disease epidemic is negligible. Watson and his colleagues (2007, 1) claim that risk of epidemics after natural disasters "is associated primarily with the size and characteristics of the population displaced, proximity of safe water and functioning latrines, nutritional status of the displaced population, level of immunity to vaccine preventable diseases such as measles, and access to health care services."

Nevertheless, the WHO warned of the potential outbreak of cholera, malaria, typhoid fever, and leishmaniasis after the 2004 Bam earthquake in Iran (Zarocostas 2004).

Moreover, much anticipated epidemics did not occur after the 2004 IOT in Indonesia, Thailand, India, and Sri Lanka – the major countries that accounted for more than 95 percent of all deaths caused by the event (WHO 2005a). Once again, the WHO issued a warning about the possibility of epidemics after the 2004 IOT: "There is an immediate increased risk of waterborne diseases, i.e., cholera, typhoid fever, shigellosis and hepatitis A and E.... Outbreaks of these diseases could occur at any moment" (WHO 2005b). In both cases, epidemics did not occur probably because many international and national NGOs responded to these events by investing their efforts; "time, personnel, and money in gearing up for potential epidemics, and considerable stocks of antimicrobial drugs, rehydration fluids for cholera patients, and vaccine were sent to the field" (Floret et al. 2006, 543).

Also, neither epidemics nor emergency health incidents occurred after the Kashmir earthquake in Pakistan, the Cyclone Nargis in Myanmar, and the Sichuan earthquake in China. Despite the presence of several thousands dead bodies in Banda Aceh in Indonesia after the 2004 IOT, no epidemic occurred among tsunami survivors. This is "the most convincing evidence to date that dead bodies pose a negligible threat to the general public after natural disasters" (Morgan et al. 2006, 813). In fact, in post-disaster period, the surviving population is much more likely to spread disease (Morgan 2006).

Note that a cholera outbreak did occur after the 2010 Haiti earthquake, but it was not due to exposure to dead bodies. Cholera was imported from Nepal by UN peacekeepers who were in Haiti for relief efforts. The Haitian cholera outbreak is believed to have started near a UN camp where Nepali peacekeepers were staying. The suspected source of Vibrio cholerae was the Artibonite River, from which most of the cholera-affected people had consumed water.

Second, exposed dead bodies, those people either killed by a major disaster or a post-disaster epidemic, are incorrectly perceived as sources of spread of disease. In reality, dead bodies have less potential to spread disease than live bodies, and therefore such bodies pose no more threat of epidemic or large-scale disease outbreak in the immediate aftermath of a natural disaster. In part, this is because infectious agents do not survive beyond 48 hours in a dead body. An exception is HIV which survives up to six days post-mortem (Morgan 2006). A decaying body is not a particularly favorable environment to harbor pathogens for diseases such as cholera, fever, typhoid, typhus, and smallpox: "…[I]f the causal agent was ever present – a rare occurrence in a normally healthy population – the microorganisms quickly cease to proliferate and progressively die off in cadavers" (de Goyet 2007, 4). This means that a human carrier of any disease is less of a health threat dead than alive. Thus, the dead in the post-disaster period do not pose any credible public health risk to the general population, except for the risk of directly contaminating drinking water supplies by fecal material released from dead bodies (de Goyet 2004).

The belief that dead bodies caused by natural disasters pose a serious health threat often leads agencies to take misguided action, such as having quick mass burials by dumping bodies in common graves without identification, which can add to the burden of suffering already experienced by surviving relatives, friends, and members of social networks and connections. These disposals create lifelong stress, hampering quality of life and even shortening the lifespan of surviving family members. Based on the myth that dead bodies caused by severe flooding from heavy rainfall from 2004 tropical storm Jeanne in Haiti, which killed more than 3,000 people, posed a high risk of epidemic outbreaks, the authorities arranged mass burials and mass cremations to dispose of bodies quickly. Not only are agencies quick to dispose of bodies; often the public initiates mass burials and cremations. Due to the unbearable smell of decomposition, the local people in Haiti soaked bodies in gasoline and burnt those on site after the 2010 earthquake (Gupta and Sadiq 2010).

To curb a potential epidemic, Thailand authorities hastily buried dead bodies after the 2004 IOT without identification of foreign tourists or local people. The same action was repeated in China after the 2008 earthquake that hit the city of Chengdu in Sichuan province, which killed 87,476 people (Gupta and Sadiq 2010). The rescue workers and first responders also sprayed bacterial solutions into the air to sanitize the environment and prevent potential contamination by dead bodies. This measure came from the common perception that the smell of decomposing human tissue must somehow be infectious. As noted, the above action was taken without respecting the process of identifying and preserving bodies, which not only goes directly against cultural norms and religious practices, but also has social, psychological, emotional, economic, and legal consequences that add to the suffering of disaster survivors. Dead bodies should be administered in a way that it is possible to identify them sooner or later.

Mass fatalities and management

A mass-fatality may be caused by natural disasters, human-related hazards (e.g., airline accidents and tunnel collapses), and pro-active human hazards (e.g., terrorist acts). Here, interest is exclusively centered on disaster-induced mass-fatalities. Before presenting definitions of mass mortality caused by natural disasters, we should note that high mortality is associated with the most major extreme events, such as the 2004 IOT, the 2005 Kashmir earthquake in Pakistan and India, the 2008 Cyclone Nargis in Myanmar and the 2008 Sichuan earthquake in China, and the 2010 Haiti earthquake. The first and last events killed more than 200,000, while the remaining disasters killed between 73,338 and 138,366 people (Gupta 2013; Gupta and Sadiq 2010). Smaller-scale or minor disasters (i.e., with death tolls of fewer than 30 people) account for only 14 per cent of total disaster mortality (UNISDR 2016). Kailash Gupta (2009) claims that disaster-induced mass-fatalities have been increasing in intensity and frequency.

There are several definitions of a mass-fatality caused by natural disasters. One is that it can be simply defined, without specifying any threshold, as the event that produces more deaths than can be handled and managed with locally available medical and other resources (McEntire 2007). Similarly, Teahen (2012, 1) defines a mass-fatality incident as "an event that causes loss of life and human suffering, which cannot be ...[addressed] through usual individual and community resources." Since "available resource" and magnitude of mass-fatality differ from community to community, the threshold of number of deaths is different therefore for each community, region, state, or country. Thus, the minimum number of deaths for a natural disaster event to be considered a mass-fatality incident varies from geographical scale of the affected area and its ability to handle the deaths. Therefore, Hurricane Katrina, which killed 1,577 people, is widely considered a mass-fatality. One issue must be noted: mass-fatalities are not the same as multiple fatalities. Although both terms are defined in a similar manner, mass casualties are different from mass-fatalities in two ways: (1) mass casualties require medical intervention, and (2) mass casualties refer to large numbers of injuries rather than many deaths (WHO 2007).

However, it is clear from the existing definitions of mass-fatalities that the number of deaths alone does not characterize the events. "If the local resources are adequate to handle the number of deceased, an incident may not be considered a mass-fatality incident" (Gupta 2013, 34). This suggests that not only does complexity increase with large numbers of deaths but also that such events require additional resources, which may not available in a given community in timely fashion. The conditions of the deaths, particularly how the deaths occurred, also determine the mass-fatality event (Jensen 2000). Thus, a mass-fatality becomes a catastrophic mass-fatality event when the number of deaths overwhelms the disaster-affected area's resources to properly handle the situation in a timely way. This type of event is closely associated with mega-disasters. In the United States, federal, state, and local governments have tool kits regarding management of mass and catastrophic fatalities. In fact, a catastrophic mass-fatality in the country is likely to trigger a disaster declaration.

Mass-fatalities caused by natural disasters involve several major steps. However, there is no agreement in the literature on the number and order of stages of dealing with the dead bodies. These stages differ by the nature, site, number of deaths, and availability of services. For example, while studying mass fatalities caused by the 2004 IOT in India and Sri Lanka, Aurther Oyola-Yemaiel and Gupta (2005) classified response to a mass-fatality into seven stages or phases: (1) body recovery, (2) transportation of body to (makeshift) morgue or local hospital, (3) preservation, (4) identification, (5) communication to surviving family members, (6) return of remains to family members for proper disposition, and (7) proper disposition of unidentified bodies by the authorities. Each of these stages must be performed in a manner that ensures the deceased are treated with respect, dignity, and in accordance with social customs and religious rituals (Jensen 2000). In most cases, government officials, non-profit organizations, and citizens themselves are faced with the difficult task of dealing with hundreds and thousands of dead bodies.

Body recovery, transportation, and preservation

Generally, mass-fatality management (MFM) begins with search and body recovery operations of human remains, which may be full body or body fragments only. This phase generally lasts several months, but most of dead bodies are recovered within first few weeks (Morgan et al. 2006). The bodies can be littered the streets, homes, on sidewalks, in open spaces, on embankments, and in ditches and other water features. Also, unlike earthquakes, in cases of tornadoes, and tropical cyclones and associated storm surges, dead bodies are found at varying distances up to 10 miles (15 km) from the places the persons originally are. In developing countries, most of the dead bodies are generally recovered by family members, relatives, friends, neighbors, and community members. For example, after the 2004 IOT in Sri Lanka, almost all bodies were recovered exclusively by the affected communities themselves (Morgan et al. 2006). In developed countries body recovery is often done by a large number of individuals, including surviving community members, volunteers of different organizations, search and rescue (SAR) teams, and military, police, or civil defense personnel.

After recovery, the remains are then placed in body bags, bed sheets, or plastic sheets, and transported to morgues in hospitals and private funeral homes to preserve them in cold storage or refrigerators for identification. Because of overflow, temporary morgue services are generally opened. No electricity after disasters makes it difficult to maintain cooler temperatures to preserve dead bodies, and lack or shortage of refrigerated containers is a serious problem in many countries. In such cases, local authorities use dry ice or temporary burial in shallow trench graves, which cannot be considered disposal of remains without identification.[14] The temperature underground is lower than at the surface, and burial acts as "natural refrigeration." Unfortunately, effective use of dry ice is often difficult to achieve because placing it on top of bodies can damage them because of its low temperature. Handling large quantities of dry ice also results in many skin burns. An effective method is to build "a small wall of dry ice surrounding a group of bodies, and then to cover the group with a tent or tarpaulin" (Morgan et al. 2006, 811).

Identification and other stages

Other activities include identification, processing, communication with surviving family members, and return of identified dead bodies to family or kin members. Before storing dead bodies, there is a need to establish scientific/forensic protocols, which are aimed to identify the victims, and to determine the time and place of death along with the cause and manner of death. The protocols also call for providing death certificates and notify the family. As a process of identification, there is a need to tag, register, and record each dead body. In Hurricanes Katrina and Gustav, RFID (radio frequency identification) tags were implemented so that later the identity of the person could be determined (Gupta and Sadiq 2010).[15]

To identify the cadaver, multiple methods are used. Simple visual identification is usually done by recovery teams and local volunteers in the first few days after the disaster before national and often foreign forensic teams arrive. This type of identification is often difficult, particularly when faces of cadaver are covered with blood and dirt. In such cases, it is possible to misidentify.[16] However, depending on the number of the dead bodies, the forensic teams can also store the bodies for later identification with modern forensic technology. The recovering teams generally identify such bodies by examining personal belongings such as money bags, identity cards, jewelry, and even mobile telephone SIM cards. These belongings should not be separated from the corresponding remains during body recovery phase. Sometimes, public authorities mandate local police, medical staff, and freelance photographers to take photographs and collect fingerprints of all recovered bodies, and record information about victims' gender, height, and age (Morgan et al. 2006).

Forensic teams identify dead bodies by analysis of DNA from bone and teeth, dental records, finger printing, and forensic pathological examination. But such identification is not available in many developing countries. For example, in the case of the 2010 Haiti earthquake, government of the country did not identify the dead bodies, except foreigners. This is because of the limited availability of forensic identification (Gupta and Sadiq 2010). In this country, family members, friends, relatives, neighbors, and passersby identified Haitian human remains (Gupta 2013). In contrast to Haiti, in Thailand, the Thai Tsunami Victim Identification (TTVI) center identified 61 percent of 2004 IOT victims using dental examinations, 19 percent using fingerprint records, 1.3 percent using DNA analysis, and 0.3 percent using physical evidence (Morgan et al. 2006).

Identifying the dead and ritual burial/cremation is essential for grieving in most cultures. However, identification is often regarded an overwhelming task to accomplish. Therefore, all dead bodies are not identified. Some bodies are so decomposed that there is no way to identify them. Additionally, hot temperature and high humidity increases the rate of decomposition: discoloration and bloating of a human face make it very difficult to visual identify dead bodies after two to four days (Morgan et al. 2006). Thus, sooner is better for victims' identification. Wrong identification is also possible. It happened in Sri Lanka after the 2004 IOT and in Haiti after the 2010 deadly earthquake. Family members of deceased get impatient and unhappy if the bodies of their loved ones are not identified. However, ultimately identified bodies are returned to the family members and they dispose by burial or cremation according to local customs and traditions.[17] Because of funeral cost, sometimes the families themselves burn or burry dead bodies of their close relatives. Bodies of foreign victims are generally repatriated by their respective embassies.

As noted, unidentified human remains are buried in mass graves or burned together. However, finding suitable government land for common graves or cemeteries is often difficult. To save land, unidentified dead bodies are buried haphazardly in several layers. Such practices need to be avoided because these obviously end the possibility of identification of human remains forever. WHO and PAHO also do not recommend common graves or mass cremation of unidentified corpses in all circumstances because these practices violate the human rights of the surviving family members and can have long-term psychological effects (PAHO 2004). If the mass burial is not avoided, PAHO also recommends burying cadavers in a way that may allow later unearthing for identification. But most mass burials are conducted by authorities in a way that prohibits any possibility of proper identification and leaves thousands of mourning survivors in misery (Gupta 2009).

Note that identified dead bodies are often buried in mass-graves in many developing countries. Figure 3.2 shows such a photo of a mass-grave of 26 Sidr victims buried together in Sarankhola, Bagerhat district in Bangladesh. As noted, a Category 4 Cyclone Sidr killed nearly 4,000 people in 2007. All of the 26 dead bodies were found by local people, including surviving family members. Because of poverty, victims were buried in one grave.

However, disposal of unidentified bodies typically is done by the national military or police, who have also legal responsibility for victim identification. Notably, after the 2004 IOT, there were 14 mass graves around Banda Aceh, Indonesia. The largest of these graves reportedly contained 60,000–70,000 victims, all of whom took some two months to bury. Moreover, two other

Figure 3.2 A mass-grave where 26 victims of Cyclone Sidr were buried in Sarankhola, Bagerhat district in Bangladesh. The cyclone made landfall on November 15, 2007. The brick walls seen in the picture were constructed later.

Source: The photo was taken in 2020 by Sarankhola Press Club President Mr Ismail Hosan Liton.

tsunami affected countries (Sri Lanka and Thailand) had to retrieve dead bodies from mass graves after receiving pressure from foreign governments for identification. Also, in the case of mass graves, mistakes can occur. For example, after the 2004 tsunami, it was suspected that some dead were buried as Muslims, when in reality, they were from Hindu and Buddhist communities (Morgan et al. 2006).[18] Figure 3.3 presents a simplified graphical depiction of the handling of deaths caused by natural disasters.

In general, public authorities provide support to surviving family members. For this purpose, family disaster management authorities will open a family assistance center (FAC). Finally, such agencies provide mental and physical support of the first responders who participate in three previous steps, particularly in developed countries. Mental or psychological support is still an alien concept in many developing countries.

The management of mass fatalities is generally supported primarily in developed countries by many public agencies and non-profit organizations, including national military, local government and police, public health personnel, including physicians, nurses and other hospitals workers, forensic

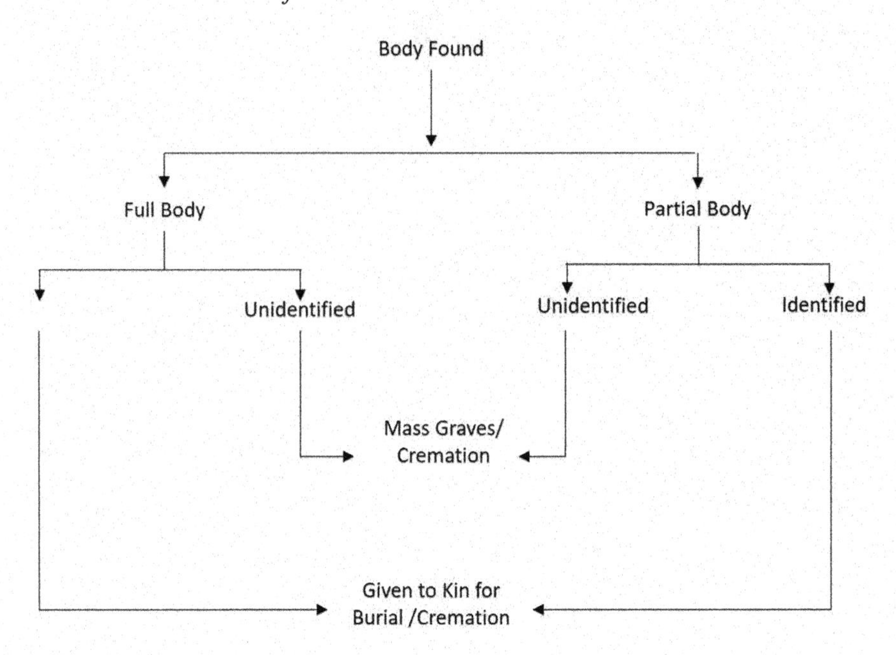

Figure 3.3 Simplified graphical depiction of dead bodies found after natural disasters.

specialists, and other individuals and volunteers. As noted, after rescuing injured persons, the SAR teams look for deceased bodies and send them to morgues nearby. Following their protocols, the forensic teams keep note of each deceased's external appearance, specific marks, and personal belongings. They are also involved in issuing death certificates and collecting post-mortem data. Photographs are taken in every case, mostly using digital cameras. Given a lack of refrigerators, the teams quickly perform these activities before the corpse is decomposed. The longer the time to decompose, the more difficult it is to identify the body. Dentists or hospitals then test DNA specimens, which are supposed to be collected from all dead bodies.

There are also potential health risks for persons handling dead bodies such as being exposed to blood, body fluids, or feces of cadavers often infected with chronic infections such as hepatitis B and C, tuberculosis, HIV/AIDS, and gastrointestinal pathogens (Morgan 2004). But these infections can be avoided by wearing heavy-duty gloves and face mask, boots, and by washing hands with soap and water.

Some individuals need psychological support because of handling so many bodies, and back injuries, caused by lifting bodies into trucks, are common among the people who are responsible for such operations. Members of body recovery teams also face potential injury risk from working among debris. Also, persons involved in one or more activities of a mass-fatality event often face heavy workloads and are under considerable stress. This, in turn, leads to a high turnover of staff. On top of that, the management of dead bodies often is severely hampered by local and national political situations (Morgan et al. 2006).

The foregoing discussion suggests that no single person or organization has a clear mandate to coordinate the process of handling all operations related to a mass-fatality. However, mismanagement of mass-fatalities following natural disasters and a chaotic way of handling the events has political and psychological implications. Sometimes these events even lead to formation of a new political system. The birth of Bangladesh, which was an eastern province of Pakistan, is a classic example. In 1970, Bhola cyclone hit East Pakistan and killed an estimated 300,000 people (Gupta 2013), but response by the Central Government of Pakistan, located in West Pakistan (now Pakistan), was grossly inadequate. As a consequence, the people of East Pakistan fought for independence, and Bangladesh was born the following year.

Mismanagement of mass-fatalities can be avoided by planning strategies to adequately handle such events in a timely way. Well laid out protocols for managing mass-fatalities are urgently needed, particularly for developing countries. In essence, there should be local and national plans for mass-fatalities as well as a clear mandate to coordinate the process of collecting, preserving, identifying, and disposing of the dead either nationally or locally. Lack of coordination causes tension, confusion, and stress not only among the individuals who handle the dead, but also surviving family members, relatives, and friend. Developing countries should emphasize forensic techniques to identify cadavers because these techniques can correctly identify decomposed and damaged bodies (Morgan et al. 2006). However, large disasters require large numbers of trained personnel. Note that DNA identification is expensive and technologically demanding; therefore, this analysis should not be considered as a first-line method of identification of dead bodies, particularly in developing countries (Morgan et al. 2006).

Conclusion

As noted, analysis of aggregate disaster deaths for the study period reveals a declining trend. This can be further decreased by improving living standards, developing resilient infrastructure, introducing early prediction and warning systems, and strengthening disaster response systems. Like aggregate disaster deaths, all natural disasters showed a declining trend, excepting earthquakes and extreme temperatures. Therefore, particular emphasis should aim to reduce earthquake- and extreme temperature-induced deaths. That cannot be possible without development and upgrading of early warning systems and strict implementation of building codes in earthquake-vulnerable regions of the world.

Tom Mitchel (2014) of the Overseas Development Institute (ODI) in London outlined seven myths about natural disasters; four of them are related to deaths (also see Jensen 2000). The first myth is that fewer people are now dying in disasters. As noted, this is proved to be true because the analysis of world disaster deaths for the period 1991–2015 shows an overall declining trend. Data available for the most recent years after 2015 also support this trend. Long-term historical data also demonstrate that the world has seen a significant reduction in disaster deaths through earlier prediction, more resilient infrastructure, emergency preparedness, and response systems. This trend is despite the one or two deadly disasters (e.g., the 2004 IOT and the 2010 Haiti earthquake) that greatly affected death figures for one or two years in a relatively long study period.

Another disaster myth is that floods and storms are the deadliest hazards. In fact, earthquakes are the deadliest, accounting for about 39 percent of all disaster deaths during the 1991–2015 period. Floods and storms accounted for about 70 percent of all recorded disaster events in the last three decades, while earthquakes were responsible for just 7 percent of the recorded events (Mitchel 2014). The third myth is that the very poorest countries experience the greatest number of deaths, which is not true. The fact is that more deaths have occurred in middle-income countries since 1980 than in low-income countries. This is because many middle-income countries (e.g., India, China, Indonesia, Thailand, the Philippines, and Mexico) are highly vulnerable to natural disasters. Also, middle-income countries have higher populations than do low-income countries. Thus, improving living standards, infrastructure and response systems in these two types of regions will be keys to preventing deaths from natural disasters in the coming decades.

The last myth is that an equal proportion of males and females die from natural disasters. The fact is that more women than men die in disasters. The typically low status of women and widespread discrimination, particularly in middle- and low-income countries, engender more female deaths. In the context of Hurricane Katrina, Jacob and his colleagues (2008) added one more myth: that natural disasters are random killers. The fact is that these extreme events more often kill the most vulnerable and marginalized groups such as minorities, the poor, the elderly, children, and women.

Among the myths related to disaster deaths, perhaps the most common myth is that disasters cause epidemics. Though people frequently believe this, it is rarely ever true (Jacob et al. 2008). Disasters rarely cause epidemics, and dead bodies do not lead to outbreaks of epidemics and infectious diseases. Such bodies pose little or no threat to public health. However, improving sanitary conditions and educating the public about hygiene measures are the best means to preventing any diseases after the disaster.

In many developing countries, mass-fatality and its management have not received much attention from disaster researchers for several reasons. They are still focusing on providing emergency relief to survivors, grossly ignoring other variables of disaster-induced mass-fatalities. Another possible reason may be because dealing with death is seen in many societies as a taboo (Bertman 1974). The way of collecting, processing, and disposing of human remains continues to be ambiguous. There is also a lack of understanding regarding religious and other cultural and social beliefs of surviving family members and others close to the family (Gupta and Sadiq 2010). These beliefs are so sensitive that "even unknowingly one may harm another" (Ekici et al. 2009, 517). Disaster managers and personnel involved in response stage of disaster management cyclone should be educated with local culture and traditions and accordingly follow in handling dead bodies and other related tasks.

Notes

1. The 2016 World Disasters Report covered the reported number of deaths for years 2006 through 2015. The 2011 World Disasters Report also published reported deaths for years 2006 through 2010. When compared these two sources of data, there are some discrepancies in the number of disaster deaths for the years 2006–2010. To maintain consistency, the deaths reported for these five years were drawn from the 2011 World Disasters Report (IFRC 2011).
2. CRED has been maintaining EM-DAT since 1988. Starting in 2014, EM-DAT also georeferenced natural disasters, adding geographical values to numeric data (CRED, USAID, and UNISDR 2015).
3. This represents 1.3 percent of global deaths (see Ritchie and Roser 2019).
4. The 2019 is the most recent year for which data on the number of disaster deaths can be available. Approximately 11,000 people died from a total of worldwide 409 natural disasters in 2019. This figure was slightly higher than the number of deaths in 2018. However, the deadliest disaster of 2019 was the flooding in India, which killed approximately 1,750 people (Statista 2020).
5. Following the 2004 IOT, a Tsunami Early Warning System was installed in the Indian Ocean, which now provides tsunami alert through three regional watch centers in Australia, India, and Indonesia. In addition, a network of 26 national tsunami information centers was also established after the 2004 IOT (CRED, USAID, and UNISDR 2016). However, none of the early warning system properly worked in 2018 when Indonesia was affected by more than one tsunami (Paul 2018).
6. Droughts are often multi-year events. For this reason, CRED has adjusted the following rules as regards mortality data for multi-year drought events: "The total number of deaths reported for a drought is divided by the number of years for which the drought persists. The resulting number is registered for each year of the drought's duration" (IFRC 2013, 226). Further some disasters begin at the end of a year and may last some weeks or months into the following year. In such cases, CRED has applied the total number of deaths for both the start year and the end year (IFRC 2013).
7. The extreme temperatures in 2010 affected an area that was about twice as large as the area affected in 2003 (European Commission 2011).
8. Only nine deaths were caused by volcanic eruptions in 2016. All deaths occurred in Indonesia (Ritchie and Roser 2019).
9. For a longer period (1966–2015), the percentages of deaths caused by the two broad types of disasters are almost similar to the percentages reported in this study. For this period, UNISDR (2016) claims that climate-related disasters

were responsible for 60 percent of all natural disaster deaths, and geophysical disasters contributed 40 percent of deaths.

10. Overall, 8,336 natural disasters occurred between 1991 and 2015. Of these 7,556 events are classified as climate-related disasters and the remaining 780 as geophysical disasters.
11. The Ring of Fire spans for nearly 25,000 miles (40,250 km), running from the southern tip of western coast of Argentina in South America, along the west coast of the United States and Canada, across the Bering Strait, down through eastern coast of Japan, and into New Zealand. Approximately 75 percent of the world's volcanoes occur within the Ring of Fire.
12. Africa is characterized by relatively low levels of seismic activity. This activity is mainly confined to the East African Rift System and the Atlas mountain region of the northwestern part of the continent. Several deadly earthquakes struck North African countries in the last century before 1991. Only two major earthquakes occurred in Africa during the study period – the 5.8 magnitude earthquake that struck Cairo, Egypt on October 12, 1992 killed more than 500 people; and the 6.8 magnitude earthquake that struck Algeria on May 21, 2003 killed more than 200 people (ICFS 2017).
13. In their review of epidemics after natural disasters, Watson et al. (2006, 2007) noted that the extreme events that do not result in displacement are rarely associated with an outbreak of epidemic disease. They further claim that the outbreak of infectious diseases in post-disaster period is associated more with the characteristics of the displaced population (size, health status, and living conditions) than to the precipitating event.
14. Dry ice should not be placed on top of the dead bodies because it damages the bodies. Depending on outside temperature, about 22 lbs (10 kg) of dry ice is needed for each body per day (Morgan 2006).
15. RFID tags are affixed to dead bodies in order to track them using an RFID reader and antenna.
16. Following the 2002 Bali bombing in Indonesia, visual identification was incorrect for about 33 percent of victims (Lain et al. 2003).
17. Some identified bodies are buried by family members but rise to the surface. In this instance, the family members are buried again.
18. Unlike Muslims, Christians, and Jewish, Hindus and Buddhists cremate their dead.

References

Bertman, S.L. 1974. Death Education in the Face of a Taboo. In *Connecting Death: A Practical Guide for the Living*. Boston, MA: Beacon Press.

Borden, K.A., and S.L. Cutter. 2008. Spatial Patterns of Natural Hazards Mortality in the United States. *International Journal of Health Geographics* 7, 64. https://doi.org/10.1186/1476-072X-7-64.

CRED (Center for Research on the Epidemiology of Disasters), USAID (United States Assistance for International Development), and UNISDR (United Nations Office for Disaster Risk Reduction). 2016. *The Human Cost of Natural Disasters 2015: A Global Perspective*. Brussels, Belgium: CRED.

Coppola, D.P. 2007. *Introduction to International Disaster Management*. Boston, MA: Elsevier.

de Goyet, C.de.V. 2000. Stop Propagating Disaster Myths. *Lancet* 356: 153–165.

de Goyet, C.de.V. 2004. Epidemics Caused by Dead Bodies: A Disaster Myth that Does Not Want to Die. *Public Health* 15(5): 297–299.

de Goyet, C.de.V. 2007. Epidemics after Natural Disasters: A Highly Contagious Myth. *Natural Hazards Observer*, January: 4–6.

Ekici, S., D. McEntire, and R. Afedzie. 2009. Transforming Debris Management: Considering New Essentials. *Disaster Prevention and Management* 18(5): 511–522.

European Commission. 2011. The Mega-Heat Wave of 2010 – Implications for the Future, 23 June (https://ec.europa.eu/environment/integration/research/newsalert/pdf/245na4_en.pdf – last accessed June 14, 2020).

Floret, N., J-F., Viel, F. Mauny, B. Hoen, and R. Piaroux. 2006. Negligible Risk for Epidemics after Geophysical Disasters. *Emerging Infectious Diseases* 12(4): 543–548.

GoN (Government of Nepal). 2015. *Nepal Earthquake 2015: Post Disaster Needs Assessment. Vol. A: Key Findings*. Kathmandu.

Gupta, K. 2009. Cross-Cultural Analysis of Response to Mass Fatalities following 2009 Cyclone Aila in Bangladesh and India. Quick Response Report #216. Hazards Center, University of Colorado at Boulder.

Gupta, K. 2013. Seeking Information after the 2010 Haiti Earthquake: A Case Study of Mass-Fatality Management. Ph.D. Dissertation, University of North Texas: Denton.

Gupta, K., and A-A. Sadiq. 2010. Response to Mass-Fatalities in the Aftermath of 2010 Haiti Earthquake. Quick Response Report # 219, Hazards Center, University of Colorado at Boulder.

ICFS (International Council for Science). 2017. *Africa Science Plan: Natural and Human-Induced Hazards and Disasters*. Pretoria.

IFRC (International Federation of Red Cross and Red Crescent Societies). 2001. *World Disasters Report 2001: Focus on Recovery*. Geneva, Switzerland: IFRC.

IFRC (International Federation of Red Cross and Red Crescent Societies). 2011. *World Disasters Report 2011: Focus on Hunger and Malnutrition*. Geneva, Switzerland: IFRC.

IFRC (International Federation of Red Cross and Red Crescent Societies). 2013. *2013 World Disasters Report: Focus on Technology and the Future of Humanitarian Action*. Geneva, Switzerland: IFRC.

IFRC (International Federation of Red Cross and Red Crescent Societies). 2016. *World Disasters Report 2016. Resilience: Saving Lives Today, Investing Tomorrow*. Geneva, Switzerland: IFRC.

Jacob, B., A.R. Mawson, M. Payton, J.C. Guigard. 2008. Disaster Mythology and Fact: Hurricane Katrina and Social Attachment. *Public Health Report* 123(5): 555–566.

Jensen, R.A. 2000. *Mass Fatality and Casualty Incidents: A Field Guide*. Boca Raton, FL: CRC Press.

Lain, R., C. Griffiths, and M. Hilton. 2003. Forensic Dental and Medical Response to the Bali Bombing. *Medical Journal of Australia* 179: 362–365.

McEntire, D. 2007. *Disaster Response and Recovery Strategies and Tactics for Resilience*. Hoboken, NJ: John Wiley & Sons.

McGrew, J.C., Jr., A.J. Lembo, Jr., and C.B. Monroe. 2014. *An Introduction to Statistical Problem Solving in Geography*. Long Grove, IL: Waveland.

Mitchel, T. 2014. Seven Myths about Disasters, November 3 (news.trust.org/item/20141103115 34-a5v2c/ – last accessed April 9, 2020).

Morgan, O. 2004. Infectious Disease Risk of Dead Bodies following Natural Disasters. *Pan American Journal of Public Health* 15(5): 307–312.

Morgan, O. (ed). 2006. *Management of Dead Bodies after Disasters: A Field Manual for First Responders*. Washington, DC: PAHO.

Morgan, O.W., O. Sribanditmongkol, C. Perera, Y. Sulasmi, D.V. Alphen, and E. Sondorp. 2006. Mass Fatality Management following the South Asian Tsunami Disaster: Case Studies in Thailand, Indonesia, and Sri Lanka. *PLoS Medicine* 3(6): 809–815.

Oyola-Yemaiel, A., and S.K. Gupta. 2005. Response of Mass Fatalities by India and Sri Lanka following the 2004 Tsunami. Phoenix: International Association of Emergency Managers. Published in CD format.

PAHO (Pan America Health Organization). 2004. *Management of Dead Bodies in Disaster Situations*. Washington, DC: PAHO.

Paul, B.K., H. Rashid, M. Shahidul, and L.M. Hunt. 2010. Cyclone Evacuation in Bangladesh: Tropical Cyclone Gorky (1991) vs. Sidr (2007). *Environmental Hazards* 9: 89–101.

Paul, B.K. 2018. Lombok Earthquakes Reveal that Indonesia's Disaster Management is Shaky. *East Asia Forum* (www.eastasiaforum.org/2018/09/22/lombok-earthquakes-reveal-that-indonesias-disaster-management-is-shaky/) (invited), 2018.

Paul, B.K. 2019. *Disaster Relief Aid + Changes & Challenges*. Gewerbestrasse, Switzerland: Palgrave Macmillan.

Paul, B.K., B. Acharya, and K. Ghimire, K. 2017. Effectiveness of Earthquakes Relief Efforts in Nepal: Opinions of the Survivors. *Natural Hazards* 85: 1169–1188.

Perera, C. 2005. After the Tsunami: Legal Implications of Mass Burials of Unidentified Victims in Sri Lanka. *PLoS* Medical 2: e185. https://doi.org/10.137/journal.pmed.0020185.

PreventionWeb. 2020. Asia – Disaster Statistics (preventionweb.net/English/countries/statistics/index_region.php?rid=4 – last accessed on April 14, 2020).

Ritchie, H., and M. Roser. 2019. Natural Disasters. Our World in Data (https://ourworldindata.org/natural-disasters – last accessed December 5, 2019).

Teahen, P.R. 2012. *Mass Fatalities: Managing the Community Response*. Boca Raton, FL: CRC Press.

Statista. 2020. Annual Number of Disaster Events Globally from 2000 to 2019 (statista.com/statistics/510959/number-of-natural-disasters-events-globally/ – last accessed on April 11, 20120).

Stott, P., D.A. Stone, and M.R. Allen. 2004. Human Contribution to European Heatwaves of 2003. *Nature* 432: 610–614.

UNDP (United Nations Development Program). 2015. *Human Development Report 2015: Work for Human Development*. New York, NY: UNDP.

UNDP (United Nations Development Program). 2016. *Human Development Report 2016: Human Development for Everyone*. New York, NY: UNDP.

UNISDR (United Nations Office for Disaster Reduction). 2016. Live to Tell: International Day for Disaster Reduction (http://disasterdoc.org/how-do-people-die-in-disasters/ – last accessed January 25, 2020).

Watson, J., M. Gayer, and M.A. Connolly. 2006. Epidemic Risk after Disasters. *Emerging Infectious Diseases* 12(9): 1468–1469.

Watson, J., M. Gayer, and M.A. Connolly. 2007. Epidemics after Natural Disasters. *Emerging Infectious Diseases* 12(1): 1–5.

WHO (World Health Organization). 2005a. Three Months after the Indian Ocean Earthquake-Tsunami. (www.who.int/hac/crises/international/asia_tsunami/3months/report/en/print.htmi – last accessed April 3, 2020).

WHO (World Health Organization). 2005b. South Asian Tsunami Situation Report 4, January 2 (www.who.int/hac/crises/international/asia_tsunami/sitrep/04/en/index.html - May 3, 2020).

WHO (World Health Organization). 2007. *Mass Casualty Management Systems: Strategies and Guidelines for Building Health Sector Capacity.* Geneva.

Zarocostas, J. 2004. Who Praises Bam Response but Warns of Disease. *Lancet* 363: 218.

4 Circumstances and causes of disaster deaths

This chapter covers the circumstances and (root) causes of disaster deaths. Circumstances of death here largely address mechanism (direct vs. indirect deaths), and location of deaths (indoor vs. outdoor, and on scene vs. hospital) (Chiu et al. 2013). They also include activity at the time of death, timing of death (day vs. night), whether the deceased person was aware of the disaster, and whether s/he attempted to take safety measures. This information does not only differ by disaster type, but also by a specific type of event occurring over time. The information is also the foundation for identifying the (medical) cause and manner of death by natural disasters. In fact, this sort of information can provide key clues to the medical examiners or coroners as they certify the death, particularly in developed countries.

Analyses of the causes of death associated with natural disasters are required to establish effective and adequate measures and interventions to reduce the extent of mortality from these extreme events. This is particularly important to the government and other agencies involved in disaster risk management (DRM). Some countries such as the Germany and United States have developed toolkits to identify disaster-specific deaths, which makes it easier to document such deaths (DKKV 2012; CDC 2017).

Cause of deaths here loosely refers to medical cause of death (e.g., trauma, drowning, or hit by moving debris). In many countries, a death certificate, which specifies the exact medical cause of each disaster-induced fatality, is not issued, while in other countries it takes a long time to issue such a certificate. As noted, in developed countries, such as the United Unites, medical examiners inspect the circumstances and evidence surrounding each death and use a case-classification method to determine whether or not the death is a result of natural disaster (Combs et al. 1996; Shultz et al. 2005). In developed countries, the cause of death can influence who is responsible legally and whether or not family members are eligible to receive federal disaster funeral assistance funding. Thus, analyses of causes, timing, and circumstances associated with disaster mortality provide valuable insights for selecting determinants of fatalities caused by extreme natural events as well as for public interventions required to lower such deaths.

Because of the complexities associated with disaster fatalities, information on circumstances and causes of disaster deaths is required to ensure the death is appropriately attributed. Additionally, this information determines the extent and scope of the health effects on the affected populations. Further, it allows experts to assess the human health impacts of a disaster and evaluate potential problems related to planning and prevention strategies for future disaster response (CDC 2017). An increased understanding of these events and their damaging effects has allowed the developed world to dramatically reduce the extent of mortality-induced deaths by natural disasters (Cardona 2004). Finally, cause of death varies across disaster types. For example, death by drowning during floods and storm surges is very common, while it is rarely the cause of death during earthquakes. For this reason, causes and circumstances of death are discussed by disaster type.

Earthquakes

The overwhelming majority of earthquake deaths are caused by building collapse, hence the popular proverb that "earthquakes do not kill people, buildings do." Indeed, structural collapse generally accounts for three-quarters of all earthquake deaths (Coburn et al. 1992; Cross 2015). This includes collapse of residential and other buildings and infrastructure (roads, highways, and bridges). The major direct cause of earthquake deaths is fall of debris from damaged or destroyed infrastructure. Earthquake deaths are also caused by injuries, which could be direct or indirect. If the death occurs during an earthquake, it is direct; if it occurs 24 hours later, it is usually considered indirect. Moreover, direct and indirect earthquake deaths differ from one event to another.[1] Bourque and her colleagues (2007) reported that in the 1989 Loma Prieta earthquake in northern California, 57 of the 60 deaths were considered direct and the remaining three indirect. These latter three persons were injured in the earthquake, but died 24 hours later.

The principal reason for the large number of earthquake fatalities caused by building collapse is poor building construction. In many earthquake-prone countries of the world, particularly in developing countries, houses are not constructed following any building code, and thus, houses collapse even with low magnitude earthquakes. Enforcing seismic construction standards requires cost, which many people in earthquake-prone areas cannot afford to bear. Thus, poor people bear the brunt of earthquake consequences.

The remaining 25 percent of all earthquake deaths are from non-structural causes and follow-on disasters. The former includes deaths due to heart attacks, cardiac arrest, suffocation, or stress associated with the event (Noji 1997) and also people who have lost their lives in the rubble. Follow-on disasters include landslides, fires, and mud-flows. Often a follow-on disaster can kill more people than does the earthquake directly. For example, most of the deaths caused by the 1923 Great Kanto Earthquake in Japan were due to fire. At the time of the earthquake, many people in the affected areas were cooking their noon meal using stoves, which generated gas explosions that caused fires, which quickly spread throughout affected areas, destroying highly combustible wood homes. Consequently, more than 130 major fires broke out across Tokyo, lasting two days and three nights. People also died from drowning when several bridges in Tokyo collapsed. To escape fire, many residents had taken shelter under these collapsed bridges (Schencking 2013; Szczepanski 2017).

Tsunamis

Some major earthquakes occur in the ocean and trigger tsunamis, which are large scale sea waves, often called "tidal waves." When these waves reach coastal areas, they can travel far inland. Although earthquake magnitude is one factor that affects tsunami generation, there are other important factors to consider. According to the United States Geological Survey (USGS), the earthquake must be a shallow marine event that displaces the seafloor. Additionally, thrust earthquakes are far more likely to generate tsunamis than are strike slip earthquakes.[2] However, in a few cases, small tsunamis have occurred from large strike slip earthquakes (greater than 8.0 magnitude on the Richter scale). For example, the 2004 Indian Ocean Tsunami (IOT) was not a thrust earthquake. Scientists suspected that it was caused by a large underwater landslide that displaced the water. Thus, tsunamis can also be produced by large scale landslides, and volcanic earthquakes and explosions. For example, in 1883, when Krakatau in Indonesia erupted, the main cause of death was tsunamis, which killed 30,000 people (Bolt 2004).

Tsunami-induced tidal waves cause floods in the coastal areas. Since 1850, tsunamis have been responsible for nearly 450,000 deaths, many in the deadly and devastating 2004 IOT. Tsunami deaths usually are the result of people being hit by floating debris and/or the impact of water. People also die in the rubble of collapsed buildings and from electrocution, fires, and explosions from gas leaks and damaged gas tanks. The violent force of a tsunami causes instant death, most commonly by drowning (Nishikiori et al. 2006). Deaths are also caused when tsunami waves recede. For example, tsunamis damage the coastal infrastructure such as sewage containment and fresh water supplies for drinking. In turn, flooding and contamination of drinking water can cause water-borne diseases such as malaria. Tsunamis do not only kill people, but also wild animals and birds. If the dead bodies are not quickly removed, they can spread infectious and other diseases. Under these conditions, people also die long after a tsunami event, and thus such deaths would be classified as delayed or indirect deaths.

Droughts

Drought is a slow-onset disaster, which lasts for years or even decades, and takes a high human toll in terms of hunger, poverty, and the perpetuation of under-development in a region (Below et al. 2007). This natural disaster is associated with widespread agricultural failures, loss of livestock, water shortages, and outbreaks of epidemic diseases. It often leads to food insecurity via a decrease in food production due to decrease in cultivated area and crop yield, and ultimately a large number of people die from starvation. In developing countries, reduced food production means high increase in prices of food grains. Decreased food production, abnormal increases in food grain prices, and non-availability of jobs reduce the food access of people of drought-affected areas (Paul 1998a). Subsequently, poor nutrition and starvation leave people in a weakened state and hence susceptible to disease, thereby increasing mortality rates.

EM-DAT shows that droughts accounted for just 2 percent of disaster deaths worldwide in the period 1994–2013 (CRED, USAID, and UNISDR 2016) or just 4 percent of weather-related disaster deaths (CRED and UNISDR 2016). However, caution should be exercised to interpret these percentages because they do not include the most indirect deaths resulted from drought-related malnutrition, undernutrition, anemia, starvation, disease, and displacement. Such indirect deaths start to occur weeks or months after the disaster is over and are often poorly documented or not counted at all. Nevertheless, this time is marked by excessive mortality rates. Due to scarcity of drinking water and because of unhygienic living conditions, people die from water-borne diseases, including malaria, and cholera, and also from measles. Women and children are primary victims of these diseases because malnutrition and undernutrition rates are high among them.

Droughts also increase the concentration of chemicals in surface waters, rapidly increasing incidence of waterborne diseases. People also suffer from many illnesses like mental and heat stress, anxiety, and asthma; additionally, excessive heat during drought particularly targets the elderly.

Floods

Before presenting causes of death, it is essential to discuss types of floods, which are generally classified into five: river, flash, coastal or tidal floods, spring/winter, and glacial lakes outburst floods (GLOFs). River floods are caused by overflow from major rivers due to heavy rainfall, while flash floods are short duration (2–4 days) and localized events that occur suddenly due to intense rain in a relatively short period of time (Paul 2002). Thirdly, coastal floods are associated with storm surges due to tropical cyclone and tsunami waves. This latter type of flood is confined to low-lying coastal areas of countries vulnerable to such events. Another type of flooding occurs in spring/winter only in areas of temperate climate due to ice jam and rapid melting of snow. This is called spring/winter flooding and is primarily driven by heavy snow years. A GLOF is a type of outburst flood that occurs suddenly when the dam containing a glacial lake fails. This type of flood is common in mountainous areas such as the Alps, the Hindu Kush and the Himalayan Mountains (Horstmann 2004).

Among these five types, flash and coastal floods cause higher average mortality both in absolute and relative terms than do river floods because of sufficient time for warning and evacuation. The first two types of floods, particularly flash floods, are relatively more rapid onset disasters than are river floods. In terms of timing, most deaths caused by flash floods occur during flood phase or impact phase, while for the other two types of floods, deaths can occur in all three stages/phases: pre-disaster stage, during disaster stage, and post-disaster or receding stage.

Irrespective of level of development and flood types, two-thirds of flood deaths are caused by drowning (Jonkman and Kelman 2005; Fitzgerald et al. 2010). A significant proportion of these deaths are caused by automobile-related in developed countries such as the United States (Chang and Chang 2019). However, the one-third of flood deaths in developed countries are caused by physical trauma, heart attack, electrocution, fire, or carbon monoxide poisoning (Jonkman and Kelman 2005; Fitzgerald et al. 2010). Physical trauma-related deaths includes those associated with people being hit by debris in the water, crashes of vehicles and collapse of buildings due to floodwater (Jonkman and Kelman 2005). In contrast, infectious diseases account for one-third of flood deaths in developing countries, but this depends on the context. Snakebite was the second largest cause of mortality in the 2007 floods in Bangladesh (Alirol et al. 2010).

Since most of the flood deaths are caused by drowning, it deserves elaboration. World Congress on Drowning (WCD) (2003) defines the drowning term as "the process of experiencing respiratory impairment from submersion/immersion in liquid." Moreover, those who die from drowning, particularly in developing countries are usually children, the elderly, and women (Telford et al. 2006). However, drowning can occur for different reasons. For example, the high number of drowning deaths during the 2011 floods in Cambodia prompted the Ministry of Health to conduct an assessment to explore the underlying causes. The National Institute of Public Health (NIPH) assessed all flood deaths in the flood-affected areas (Saulnier et al. 2019) and reported that 47 percent of all drownings were caused by fall into water while adults were performing daily work, such as fishing, taking care of their rice fields, or finding grass for their livestock. As in other flood-prone developing countries, people in Cambodia travel by boats during flood season, and thus capsizing boats caused 18 percent of drowning deaths in 2011.

The NIPH further reported that less than 50 percent of the drowning deaths occurred in open or permanent water, like lakes, or rivers. Additionally, children under the age of 15 years accounted for 39 percent of all drownings. They fell into water from raised houses, especially during the day when parents were gone and children were alone.

> Eighty percent of the deaths coincided with the peak of the flood, while another 15 percent happened during the initial rise of water before the flood and during the recession of the flood water. Drownings were most

> frequent when flood waters reached villages and rice fields, and the two main locations for drowning were in flooded rice fields, or in flood water under houses.
>
> *(Saulnier et al. 2019, 18)*

In Cambodia, many houses, particularly near a major river, are built on a platform.

Assessing 13 flood events in four European countries (Czech Republic, France, Poland, and the United Kingdom) and the United States between 1989 and 2003, Jonkman and Kelman (2005) analyzed the medical causes and circumstances of deaths, including timing. They classified 247 flood deaths into eight groups: drowning (67.6 percent), physical trauma (11.7 percent), heart attack (5.7 percent), electrocution (2.8 percent), carbon monoxide poisoning (0.8 percent), fire (3.6 percent), other (1.2 percent), and unknown or not reported (6.5 percent). Vehicle-related drowning occurred most frequently, mainly when people tried to drive across flooded bridges, roads, or streams. Notably, drowning more frequently occurred in the United States than in Europe. Jonkman and Kelman (2005) identified five possible reasons: (1) a better understanding about the dangers posed by flooding in Europe; (2) better warning systems and higher compliance with such warnings in Europe; (3) the nature of flooding differs between the USA and Europe; (4) fewer low-water crossings in Europe because of different road networks and flood characteristics (e.g., frequency and intensity); and (5) different reporting systems for flood deaths.

However, 14 deaths occurred from heart attack, three of which happened during the pre-flood stage of the evacuation process. Jonkman and Kelman (2005) further reported that all carbon monoxide poisoning occurred in buildings and during a flood, fire can start from faulty electric connections and/or candles. All reported flood deaths caused by electrocution and fire occurred in the United States.

Jonkman and Kelman (2005) disaggregated circumstances of deaths caused by drowning and physical trauma into five groups: as a pedestrian (26.7 percent), in a vehicle (38.5 percent), after a fall from a boat (3.6 percent), during a rescue attempt (1.2 percent), and in a building (9.3 percent). Accordingly, 196 of the 247 flood deaths were caused by drownings and physical trauma. In particular, drowning by vehicle is caused either by rapid flow of flash floodwater or car crashes. These crashes occurred when people drove over flooded bridges.[3] Additionally, three people died during the rescue process, one of whom was a rescuer.[4]

Both direct and indirect causes of flood death differ between developed and developing countries. For instance, most of the flood fatalities in developed countries are direct deaths and primarily caused by flash floods. About 66 percent of flood deaths in developed countries are caused by vehicle-related drowning, and carbon monoxide poisoning, which are more common in developed countries than in developing countries. On the other hand, snake bite is a more common cause of flood deaths in developing countries, while it is rare in developed countries. Also, most flood deaths are direct deaths in developed countries, while indirect flood deaths constitute a large proportion of deaths in developing countries.

The majority of indirect flood deaths in developing countries are caused by water-borne diseases such as diarrhea, cholera, malaria, and respiratory illnesses (e.g., cough, sneezing, and sore throat). These deaths generally occur after a flood recedes when both children and the elderly are more at risk of death from water-borne and respiratory diseases than are other people. These diseases are caused or exacerbated by lack of nutrition and pure drinking water, by the unsafe ways drinking water is stored and handled, by poor hygiene, and often by the partial and/or total deterioration of sewage and sanitation facilities (Kunii et al. 2002; Ahern et al. 2005). When floods cause a relatively large number of deaths, indirect deaths and water-borne diseases account for significant number particularly in developing countries. For example, during the 1972–2019 period, Bangladesh experienced two deadly floods, one in 1987 and one in 1988. These floods killed 2,680 and 2,440 people, respectively.[5] Both years, many people in the affected areas suffered from flood-related illnesses. This seems to be one of the main reasons for the very high flood fatalities in 1987 and 1988 (Padli et al. 2013).

Cyclones/hurricanes/typhoons

Both in developed and developing countries, storm surges are the main cause of death during a cyclone/hurricane/typhoon disaster, due to drowning. In fact, roughly half of all deaths in the United States from tropical cyclones are due to storm surge.[6] People also die from drowning not related to storm surges, but the percentages differ from year to year as well as from country to country. Edward Rappaport (2014) identified the causes of 2,170 hurricane-induced deaths directly attributed to Atlantic tropical cyclones in the continental United States for the 50-year period from 1963 to 2012.[7] He reported that 88 percent of the deaths occurred in water-related incidents, most by drowning. After storm surges, which accounted for 49 percent of all hurricane deaths, 27 percent of fatalities were caused by rainfall-induced freshwater floods and mudslides or landslides (Figure 4.1). Note that heavy rainfall and tornadoes are frequently accompanied with a hurricane. Drowning motor vehicles is the main cause of death from this type of flooding. The remaining 12 percent died because of strong wind and tornadoes associated with hurricanes, falling trees, in fires, and from carbon monoxide poisoning.

Rappaport (2014) further reported that 6 percent of people perished near the shoreline from rip currents and large waves during hurricanes in the United States. Another 6 percent died from drowning offshore within 50 miles (75 km) of the coast. Wind associated with hurricanes also killed 8 percent of the victims due to trauma from falling trees, structures, or

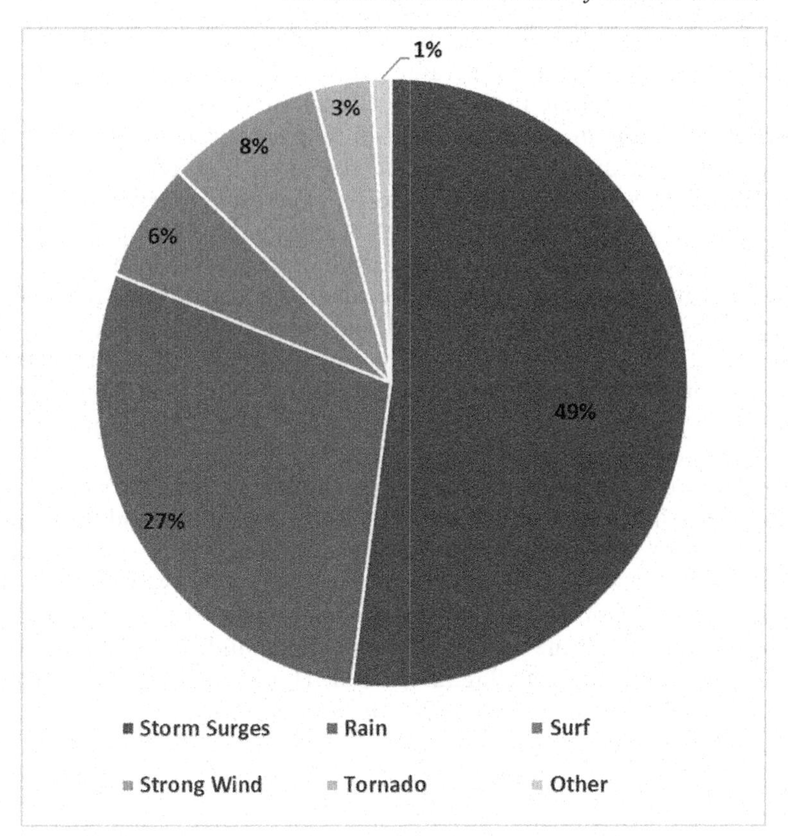

Figure 4.1 Causes of death in the United States from Atlantic Hurricanes, 1963–2012.

Source: Rappaport (2014), p. 341.

wind-propelled debris. As is heavy rain, hurricanes are also associated with tornadoes in affected areas. Hurricane-induced tornadoes killed 3 percent of all victims during the period under consideration (Figure 4.1).

Rappaport and Blanchard (2016) also analyzed indirect deaths associated with Atlantic tropical cyclones for the same 50-year period. The number of indirect deaths (1,418) is almost as large as the number of direct deaths. These indirect deaths were caused by a total of 59 storms. By far the largest number of indirect deaths was the more than 500 associated with Katrina. In about half of the storms, indirect deaths accounted for more casualties than direct deaths.[8] They classified four major causes of indirect deaths: power problems, cardiovascular failure, evacuation, and vehicular incidents. Power problems, usually in the form of loss of electricity, triggered a variety of equipment failures or other actions that contributed to loss of life. Majority of the deaths related to power problems were due to carbon monoxide poisoning, which resulted from using generators needed to regulate air temperature in the houses immediately after each the hurricane. The hurricane survivors needed generators because of loss of electricity.

Cardiovascular failure in the form of heart attacks was the most pervasive elements in indirect deaths, contributing about one third of deaths. As indicated, cardiac arrests occurred before, during, and after storms. After hurricanes, cleaning rubbles and fixing houses contributed physical (over) exertion, which ultimately caused heart attacks. The third reason for indirect hurricane deaths is associated with evacuation, which is generally occurred before a disaster. Vehicular accidents made up the fourth and final category of cause associated with hurricane-induced hurricane deaths in the United States. It accounted for more than 250 indirect deaths. All the major causes of indirect hurricane deaths occurred in combination. For example, vehicle accidents could occur while people were going to hurricane shelters. Or die due to heart failure at such shelters.

However, considering temporal trends of the causes of hurricane mortality in the United States as reported by Rappaport (2014) and Rappaport and Blanchard (2016), causes of death may differ for a particular event. Diakakis et al. (2015) analyzed the causes and the activity of the deceased at the time of death, i.e., circumstances under which each fatal incident occurred, for Hurricane Sandy, which made landfall in eight countries, six in the Caribbean Sea (Bahama, Cuba, Haiti, Jamaica, Puerto Rico, Dominican Republic) and two in North America (Canada and the United States) starting on October 24, 2012. It killed 233 people, as illustrated in Figure 4.2 by country. Of the total deaths, Diakakis et al. (2015) identified the temporal distribution of 181 deaths. They reported that the majority of the deaths occurred between October 24 and 31, 2012, and a significant number of fatalities occurred on the day of each landfall, i.e., on

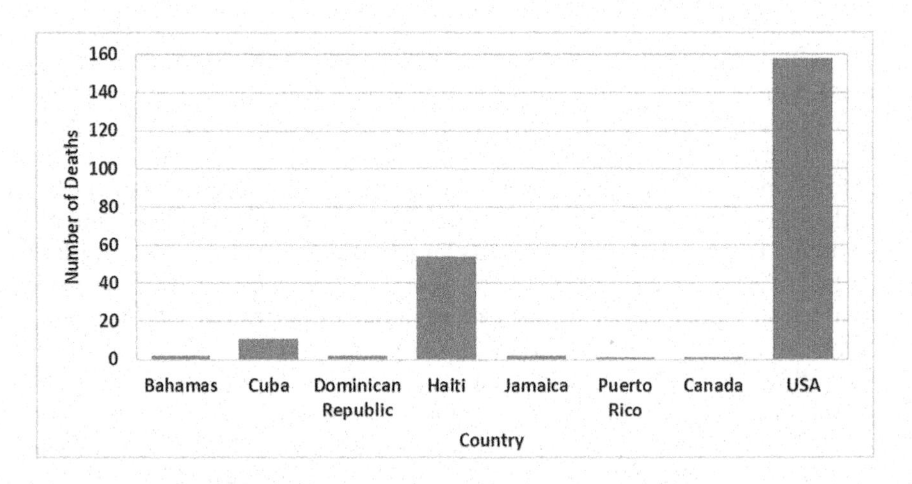

Figure 4.2 The number of deaths induced by Hurricane Sandy by affected countries.

Source: Diakakis et al. (2015), Table 1, p. 135.

October 24 and 25 October in the Caribbean and on October 29 in the United States.

All deaths associated with Super Hurricane Sandy occurred in various affected countries along the hurricane path from October 24 to November 28, 2012, long after the hurricane dissipated on November 2, 2012. Within the United States, the majority of deaths were recorded in at least seven eastern states (New Jersey, New York, Connecticut, Delaware, Rhode Island, Pennsylvania, and West Virginia). Of these states, with 71 fatalities, New York State recorded the largest number of deaths, followed by New Jersey (43), and Pennsylvania (15). Diakakis et al. (2015) reported that approximately 50 percent of deaths occurred within less than 1.5 miles (2 km) of the coast.

Table 4.1 presents information on causes of death and the activity of deceased persons at the time death. The table suggests that most deaths (37.3 percent) were caused by physical trauma (crush/cut/struck), followed by drowning, poisoning, hypothermia, heart attack, asphyxiation, electrocution, and burns. A significant proportion of deaths were categorized as poisoning caused by carbon monoxide after people left their generators running indoors. However, the cause of death for 46 of the 233 fatalities

Table 4.1 Causes of deaths and activity at the time of deaths caused by super Hurricane Sandy ($N = 233$)

Cause of death	*Number (percentage)*
Physical trauma	87 (37.3)
Drowning	51 (21.9)
Poisoning	15 (6.4)
Hypothermia	10 (4.3)
Heart attack	7 (3.0)
Asphyxiation	6 (2.6)
Electrocution	6 (2.6)
Burns	5 (2.2)
Not reported	46 (19.7)
Activity	*Number (percentage)*
Stayed indoors	97 (41.6)
Involved in clean-up	20 (8.7)
Driving a vehicle	16 (6.0)
Walking	14 (5.9)
Evacuating	6 (2.6)
Repairing	4 (1.7)
Boating	2 (0.9)
Rescuing	1 (0.4)
Not reported	68 (29.1)

Source: Compiled from Diakakis et al. (2015), 140.

was not reported (Diakakis et al. 2015). Meanwhile, a considerable proportion of the drowning deaths occurred at homes and in cars. Matt Daniel (2012) also mentioned other causes of Hurricane Sandy deaths in the United States, which are not reported by Diakakis; probably they were considered not worth reporting because of their low numbers (Table 4.1). These other causes were trees falling into homes and cars (mainly inland), and falls from the tops of stairs due to no power and having very little light to navigate in homes. Daniel (2012) also noted that people died from not receiving medical aid fast enough to prevent loss of blood or oxygen (also see CDC 2013).

With regard to causes of deaths due to Hurricane Sandy, there are several differences between those in the Caribbean and those in continental North America. More deaths were caused by drowning and physical trauma in the countries of the Caribbean relative to those in continental North America. Diakakis et al. (2015) claimed that drowning and trauma accounted for 57.1 and 28.6 percent of local deaths in the Caribbean, while the corresponding percentages for continental North America were 24 percent and 45.2 percent, respectively. Additionally, more deaths were caused by burns, electrocutions, heart attacks, hypothermia, and carbon monoxide poisoning in continental North America than in the Caribbean countries (Diakakis et al. 2015).

For fatalities due to Hurricane Sandy, the deceased were involved in nine different types of activities (Table 4.1), and information regarding the activities was available for 165 cases (70.3 percent). Accordingly, the highest percentage of deaths (41.6 percent) occurred indoors, followed by deaths of individuals involved in clean-up operations post-disaster.[9] Diakakis et al. (2015) stated that deaths that occurred indoors were dominated by people whose average age was 60.7 years old, while outdoor deaths recorded an average of 47.1 years old. More elderly people died probably because they were reluctant to evacuate. People also died while driving cars and walking on the street during the event. Other activities included evacuating a location, attempting to repair houses, doing sports, being on a boat, and attempting a rescue (Table 4.1). Diakakis et al. (2015) further reported that more outdoor fatalities occurred in Caribbean countries than the United States and Canada. In fact, ratio of indoor to outdoor deaths was 1:1.2 in the case of Caribbean countries, and for the USA and Canada, the ratio was 1:0.8.

Among the 233 deaths, direct and indirect causes of fatalities were identified for 193 or 82.8 percent of all deceased persons: 106 (55 percent) were direct and 87 (45 percent) were indirect deaths. Among direct deaths, the highest number (40 or 37.7 percent) was by drowning in the storm surge water. The other reasons in descending order are as follows: tree fall caused by hurricane wind (24 or 22.6 percent), freshwater flooding (12 or 13.3 percent), building collapse (12 or 11.3 percent), rough seas (4 or 3.8 percent), landslides (3 or 2.8 percent), wind gusts (3 or 2.8 percent), and trauma by airborne debris (2 or 1.9 percent) (Diakakis et al. 2015).

Among indirect deaths, the highest number (46 or 52.8 percent) was from causes associated with power outages, followed by car crashes (15 or 17.2 percent), tree or limb falls during clean-up (8 or 9.2 percent), heart attack (7 or 8.1 percent), contact with power lines (5 or 5.7 percent), other accidents (5 or 5.7 percent), and suicide (1 or 1.2 percent). Other deaths were due to misuse of heavy equipment during or after the hurricane. Clearly, direct deaths from Super Hurricane Sandy seem to have resulted from the forces of the hurricane, such as wind force and storm surges. These deaths occurred mostly on the days that the hurricane made landfall in different countries of the Caribbean and of continental North America. Contrarily, indirect deaths occurred most often before and after landfall days (Diakakis et al. 2015).

As in developed countries, the major cause of cyclone/hurricane/typhoon-related deaths in developing countries is due to drowning. Other causes are also more prominent in such regions than in developed countries. More people die from falling trees, lightning, and collapsing houses and shelters (Gupta 2009; Paul and Dutt 2010). People also die because of heart attacks. In developing countries such as Bangladesh, after hearing a cyclone warning, people do not go to the cyclone shelters in large numbers. Instead, some of them stay home and opt for alternative emergency measures, such as taking shelter on the roof of their residence or climbing a strong tree and tying themselves to it. However, the force of wind often damages or destroys their houses and uproots trees, and thus they either are directly killed, or severely injured. Some of the injured people later die.

As noted in Chapter 2, the Philippines hit by Typhoon Haiyan in 2013 and killed 6,300 people. There, the majority of the deaths were caused by drowning in storm surge and flood water. But a significant number of deaths were caused by building and shelter collapse. The typhoon destroyed between 70 and 80 percent of the buildings in the affected areas and subsequently many deaths (Mersereau 2013). Although Tacloban city, which was the worst affected, opened 23 designated shelters, the evacuation rate was as low as seven percent. Moreover, several of these shelters could not withstand the storm and collapsed, causing many deaths. The metal sheet roofs of most of the shelters were not tightly tied to the walls, and because of this, several shelters were blown away by strong winds. Many shelters were also located in low-lying areas, and thus were vulnerable to inundation from storm surges. In fact, of 300 people who sought safety in one of the shelters in Tacloban, at least 23 died by drowning when storm surge water inundated it (CRS 2014; McPherson et al. 2015).

Storm surges inundate coastal areas with saline water. These surges and heavy rainfall associated with a tropical cyclone often together contaminate any sewerage system. Thus, cyclone-affected areas suffer from waterborne diseases due to lack of safe water supply and proper sanitation. This was the case after Cyclone Gorky hit Bangladesh in 1991 where nearly 7,000 people died from diarrhea, dysentery, and respiratory illness during the post-disaster period (Haque and Blair 1992).[10] Deaths from water-borne diseases are more common in developing countries than in developed countries, and by definition, these deaths are indirect deaths.

Most cyclone/hurricane/typhoon-related deaths occur on the landfall day as the following data confirm. Bourque et al. (2007) reported that several hurricanes (Elena, Gloria, Hugo, Andrew, Marlyn, Opal, Georges, Floyd, and Isabel) that occurred in the United States between 1985 and 2003 killed 208 people. Centers for Disease Control and Prevention (CDC) assigned a time of death for 103 of 208 victims. Nine occurred before hurricane landfall, presumably from heart attacks, 57 during, and 37 after the hurricane. As recently as November 9, 2019, Cyclone Bulbul hit Bangladesh and killed eight people. Six of the eight died on the landfall day and the other two died a day after the landfall.

Tornadoes

A careful review of relevant literature suggests that the majority of tornado fatalities both in developed and developing countries are direct deaths and caused by trauma or trauma-related incidents. Flying objects in high winds are the main source of trauma and these objects directly kill people by hitting or cutting their bodies (Brown et al. 2002; CDC 2012; Chiu et al. 2013). People are also killed when wind throw them into hard objects. High wind throws not only people, but also other objects like cars. Thus, people are also killed by car crush during tornadoes and because of severe damage to or destruction of houses/buildings where they shelter. More than one cause can also be associated with direct or indirect tornado deaths. To clarify the causes and circumstances of tornado deaths, three tornado events are discussed below. Because the United States accounts for the vast majority of tornadoes in the world, two of these events of particular historical significance occurred in the United States. On average, the country experiences more than 1,000 tornadoes per year.

2011 Tornado Outbreak, Alabama, April 27

From April 25 to 29, 2011, the southeastern United States experienced 351 tornadoes that resulted in 338 deaths in Alabama, Arkansas, Georgia, Mississippi, and Tennessee. The deadliest day was April 27, 2011 when Alabama was hit by a record number of 62 tornadoes, killing 253 people of all ages and walks of life. Of the 62, eight were EF-4 and three were EF-5 tornadoes. Based on death certificates, Chiu and her co-authors (2013) examined the causes and circumstances of death surrounding 247 of the 253 fatalities. They attributed 235 (or 95.1 percent of the 247 deaths directly to a tornado, and the remaining 12 (4.9 percent) deaths indirectly to a tornado. They defined direct tornado deaths as being caused by the physical forces of the disaster, such as strong wind, or by the direct consequence of these forces such as flying objects. An indirect death is defined as follows: "if it was caused by unsafe or unhealthy conditions generated by the disaster (e.g., hazardous roads) or a loss or disruption of usual services (e.g., a power outage)" (Chiu et al. 2013, e3).

Of the 235 direct deaths, 122 people (47.7 percent) were struck or cut by debris or objects, 81 (34.5 percent) were thrown by wind force, 44 (18.7 percent were crushed, 4 (1.7 percent) were trapped in rubble, 59 (25.1 percent) were subject to multiple injuries, and 54 (23 percent) died from reasons not known at the time of the research. Of the 12 indirect deaths, seven (58.3 percent) died because of a power outage, including four by house fire, two deaths related to medical conditions (e.g., lack of refrigeration for insulin), and one as the result of a fall during the blackout. The remaining five deaths were caused by complications of post-tornado injury, myocardial infarction, and premature birth. Chiu et al. (2013) further reported that 223 (93.3 percent) of all deaths were caused by trauma, and 17 (6.9 percent) were trauma-related. Seven (2.8 percent) died from smoke inhalation, acute myocardial infection, Alzheimer's disease, or diabetic ketoacidosis. At least 55 (22.3 percent) persons sustained some form of head injury, which was listed as the official cause of death for 39 (15.8 percent) persons. The majority of the fatalities (212 or 85.8 percent) occurred on the day of the tornado (Chiu et al. 2013).

Of the 247 total tornado deaths in Alabama, 228 (92.3 percent) occurred inside of buildings. Among these deaths, 133 (53.8 percent) occurred in single-family homes, 51 (20.6 percent) occurred in mobile homes, 15 (6.1 percent) in apartments, and three (1.2 percent) in church, factory and hospice care facilities (Table 4.2). Of the deceased, 165 (66.8 percent) were in homes that were completely destroyed by the tornado. Chiu et al. (2013) reported that 186 (75.3 percent) individuals were in their own home, while 36 (14.6 percent) were in the homes of their relatives, friends, or neighbors, and four (1.6 percent) died at work. Fifteen (6.1 percent) persons died outdoors

Table 4.2 Summary of tornado deaths by location in Alabama and Joplin, Missouri

	Alabama	*Joplin*
Location	*Number (percent)*	*Number (percent)*
Indoor	202 (81.78)	135 (83.85)
Residential Building	199 (80.57)	80 (49.69)
Non-residential building	3 (1.21)	55 (34.16)
Outdoor	15 (6.07)	20 (12.42)
Vehicle	11 (4.45)	15 (9.32)
Open	4 (1.62)	5 (3.10)
Unknown	30 (12.15)	6 (3.73)
Total	247 (100.00)	161 (100.00)

Sources: Compiled from Chiu et al. (2013) and Kuligowski et al. (2013).

at the time of the tornado. Of them, 11 (4.5) died in vehicles. The locations of the remaining deaths are not known (Chiu et al. 2013).

2011 Joplin Tornado

An EF5 tornado tore through a densely populated section of Joplin, Missouri, on the evening of 22 May 2011, killing 161 people.[11] According to information contained in death certificates, nearly 96 percent (or 155 out of 161) of all fatalities were caused by impact-related factors (e.g., multiple blunt force traumas to the persons struck by debris during the storm). The remaining fatalities were caused by stress-induced heart attacks, pneumonia, or chronic obstructive pulmonary disease due to blunt-force trauma to the body. As of May 2013, the City of Joplin listed 158 of the 161 deaths as being direct (Kuligowski et al. 2013).

Of those 161 deaths, 135 (or 83.9 percent) occurred inside of buildings (Table 4.2), which have been disaggregated into two types: residential and non-residential. The former included apartment buildings, single residences, nursing homes, medical hospitals and clinics, and churches, while the latter included Joplin businesses, such as retail stores (e.g., Home Depot, AT&T store, and Walmart) and restaurants (e.g., Pizza Hut and Golden Corral) as well as the Elks Lodge where four people died. Surprisingly no deaths occurred in mobile homes. As in other years, mobile homes accounted for 20 percent of all tornado deaths in the United States in 2011. However, information collected from the City of Joplin revealed that out of a total 21,362 housing units, 350 (2 percent) were mobile homes in 2009, which was much lower than the national percentage. Moreover, all of the mobile home units in Joplin were outside the tornado path (Paul and Stimers 2012).

Twenty (or 12.4 percent) of the 161 deaths in Joplin occurred outside of residential and business buildings. These were individuals who came to the city to visit friends, relatives, or to shop in the mall or business stores, including restaurants. These individuals also included those who were in transit, stopped in vehicles, or caught outdoors when the tornado struck. Most of the outside deaths (75 percent) occurred in vehicles. Kuligowski et al. (2013) reported that eight deaths took place in churches, and three deaths occurred in Joplin's Stained Glass theater. Finally, 16 deaths occurred in St. John's Regional Medical Center (SJRMC) and Meadows health care facility.

Paul and Stimers (2014) analyzed the 161 deaths caused by the 2011 Joplin tornado by type of resident: residents of Joplin and those of neighboring communities. They further divided Joplin residents into two categories: residents of the tornado-affected area or damaged area and those of non-affected areas of Joplin. Of all the deaths caused by the tornado, 137 (85 percent) were residents of Joplin. Of these 137, 11 (8 percent) actually lived in Joplin but outside the damage zone at the time of the tornado. However, because the tornado occurred on a Sunday, all 11 were visiting the homes of friends and relatives located in the damage zone, attending churches, shopping mall and other stores, or dining out. Twenty-four (15 percent) people died who came to Joplin from neighboring communities for some of the reasons stated above. They were residents of 14 neighboring communities in Missouri, Kansas, and Oklahoma. The relatively large number of non-Joplin-resident deaths reflects Joplin's status as a major regional center. Joplin city lies near the borders of Missouri, Kansas, Oklahoma, and Arkansas (Paul and Stimers 2014).

Of the 161 deaths caused by the 2011 Joplin tornado, 124 (77 percent) died on the day of the event, i.e., on May 22, 2011. All deaths that were attributed to an outside location, including to a vehicle, also occurred on May 22. Most of the delayed deaths were caused by injuries, and a significant number of people died at the location where they were injured. Some injured persons also died later elsewhere. For example, six persons injured at the Greenbriar Nursing Home later died in a hospital where they were transported for urgent treatment. Kuligowski et al. (2013) reported that one of those six fatalities occurred four days after the tornado, and four died in June 2011. The two persons who were fatally injured at the Meadows Healthcare facility both died two days after the tornado hit. Five patients died at SJRMC during or immediately after the tornado. In addition, nine persons were injured at this facility, but died in other locations days, weeks, or months later (Kuligowski et al. 2013; also see Kuligowski 2020).

2005 Bangladesh Tornado

Apart from the United States, tornadoes occur in many countries in Asia, Europe, and Latin America. Among these countries, Bangladesh, one of the most tornado-prone countries, experienced its deadliest tornado on April 26, 1989, which killed 1,300 people in the Manikganj district.[12] This huge number of deaths was caused by high population density, poor construction, and lack of tornado forecasting and warnings. Another deadly tornado hit on May 13, 1996, in the Tangail district and killed 558 people (Paul 1998b). However, as for other developing countries, the causes of Bangladesh tornado deaths are rarely studied. For instance, only one study has systematically and rigorously analyzed causes of deaths for the March 20, 2005 tornado that hit two districts (Gaibanda and Rangpur) in northern Bangladesh (Sugimoto et al. 2011).

The 2005 northern Bangladesh tornado killed 56 people. Forty-three deaths (77 percent) occurred indoors, while the remaining 13 (23 percent) deaths occurred outdoors. Sugimoto et al. (2011) reported those tornado deaths by proximate and underlying causes.[13] The most proximate cause of death was a head injury (36 percent), followed by hemorrhage (25 percent), and sepsis (18 percent). The remaining 21 percent died from heart failure, chest injury, suffocation, and gangrene. Underlying major causes of deaths were as follows: multiple injuries (40 percent), hemorrhage (29 percent), head injury

(8 percent), sepsis (8 percent), chest injury (5 percent), cardio-vascular disease (CVD), back pain (3 percent), and others (3 percent). The others included shock and asthma (Sugimoto et al. 2011).

Lightning

Lightning is perceived as a lesser threat to human life than are other weather-related disasters. In reality, this event is highly dangerous and kills thousands of people world-wide (Mulder et al. 2012; Salerno et al. 2012; Singh et al. 2017). In the United States, lightning strikes cause more deaths and injuries than other natural disasters, such as hurricanes, tornadoes, volcanoes, and floods (Seidl 2006). According to the National Weather Service (NWS) Storm Data, over the last 30 years (1989–2018), the U.S. has averaged 43 reported lightning fatalities per year. Only about 10 percent of people who are struck by lightning are killed, leaving 90 percent surviving, but with various degrees of disability and often with long-lasting neurological damage. Thus lightning injures far more people than it kills.

Additionally, lightning is very common in many tropical and subtropical developing countries, particularly during rainy season. The ratio of injuries to deaths is higher in these countries primarily than non-tropical countries because of two factors: higher lightning density, and lightning-unsafe housing. Some countries account for nearly 80 percent of global lightning flashes: Argentina, Bangladesh, China, Democratic Republic of Congo (DRC), India, Indonesia, Malawi, Malaysia, Pakistan, Sri Lanka, Swaziland, and Uganda. The countries of Central Africa have a greater incidence of lightning strikes than does any other world region.

Lightning injuries and deaths are classified as direct strikes, contract strikes, upward streamer, side splash, and ground current. These terms are associated with the ways lightning strikes people. Lightning that hits someone directly from the sky is called a direct strike, and it mostly occurs to victims who are in open areas. Contact strikes are caused by touching an object that is struck by lightning. Such strikes can occur both inside and outside a building. However, direct and contact strikes each account for only 3–5 percent of lightning deaths and injuries (Williams 2013) and are the least common. The third most common cause of lighting deaths and injuries are the upward leaders or streamers that rise from high objects and the ground before lightning strikes. Upward leaders generally cause 10–15 percent of all deaths and injuries (Holle 2016).

The second most common kind of lightning strike is called "side splash" or "side flash," which accounts for 30–35 percent of lightning deaths. This occurs when lightning strikes a taller object near the victim and a portion of the current jumps from the object to the victims. Side flashes generally occur when the victim nears an object as it is struck. Finally, ground current, which occurs when lightning strikes an object on the ground and much of its energy spreads outward from the strikes in and along the ground surface, is the most common kind of strike. In this case, the distance from the lightning strike to the object is an important determinant for both deaths and injuries. Ground currents are especially dangerous to farm animals because the current passes through the entire body between the front and rear legs. Ultimately, ground current accounts for 50–55 percent of lightning-induced deaths and injuries (Williams 2013).

Most deaths by lightning strike occur either because of primary cardiac arrest (heart stopping) or hypoxia-induced secondary cardiac arrest. Some lightning victims may undergo delayed death if they suffer severe injury, including burns. In a study of injury and death from lightning in Northern Malawi, Salerno et al. (2012) found 450 victims of lightning between 1979 and 2010. Of these victims, 117 (26 percent) died. This mortality rate is substantially higher than the 10 percent rate reported earlier, which is not unusual because Northern Malawi is considered a global "hot spot" for lightning flash density.

Salerno et al. (2012) analyzed circumstances of death and injury and reported that probability of death was not different for victims who were barefoot versus those who wore sandals or shoes. Likewise, the probability of death was similar regardless of the presence of rains or not. But they did find that mortality rates were high among victims struck outside in the open or outside under cover. Furthermore, those in the open experienced the greatest risk of death (34 percent), followed by those outside under cover (28 percent), then those indoor under thatched (24 percent), and finally those indoor under tin (18 percent).

Heat waves

Both tropical and subtropical countries are prone to heat waves caused by extreme summer temperatures. Among such countries, heat waves are more frequent in the United States where this disaster has killed more people than all other weather-related natural disasters combined from 1900 to 2018 (Statista 2019). Health effects of a heat wave result from high temperatures, high humidity, lack of air movement, and radiation. According to Amadeo (2019), deaths caused by heat waves occur in four ways. First, heat stress causes loss of body salt and dehydration, which are responsible for changing in body chemistry. Second, when the body tries to cool itself, this taxes the heart. This event leads to heart failure and ultimately to death. Third, when the body temperature rises above 104°F (40°C), human organs start to fail, which causes heat stroke or sun stroke – the most serious form of heat injury. This directly causes deaths from hyperthermia. Finally, people have drowned while trying to cool off in rivers and lakes.

Heat waves are most likely to kill people who work or live outdoors such as construction workers and farmers. In both developed and developing countries, those at risk of death from heat waves are those without air conditioning and/or suffering widespread power outage due to increased demand for air conditioning. The elderly, children younger than five, the poor, and those with chronic health conditions are at high risk for deaths from severe heat waves. Generally, elderly women are at greater risk than elderly men (Bourque et al. 2007). Nearly three-fourths of heat-related deaths occur in persons aged 65 or older. However, there are no criteria to determine whether death is actually caused by heat waves because deaths are not usually directly attributable to heat waves. Soaring temperatures simply turn pre-existing conditions such as heart problems or lung disease lethal. Therefore, heat wave deaths are likely to be over or under estimated.

Ice and snow storms

In non-tropical countries, a considerable number of deaths are caused by extreme winter weather conditions such as ice storms and snow storms or blizzards. The leading cause of death during winter storms is automobile accidents. This is because heavy snow and high winds accompanying snow can create hazardous driving conditions on highways and bridges leading to many traffic accidents. In this type of circumstance, road crews find it difficult to keep highways clear. About 70 percent of winter-related deaths occur in automobiles, and about 30 percent are for other reasons. Mersereau (2016) reported that winter storms directly claimed 571 lives between 1996 and 2011. When plane crashes due to winter storms were added, this number jumps to a staggering 13,852 deaths worldwide over the same period.

Apart from car accidents, Mersereau (2016) listed six other reasons as causes of deaths during winter storms. Falling and slipping on an icy sidewalk or driveway often directly kill people or does so indirectly through severe injuries. People also die from heart attacks when clearing a driveway or sidewalk with shovels. Also, without proper clothes while shoveling, people can die from hypothermia; moreover, prolonged exposure to cold weather, such as falling into water after walking on thin ice, can dangerously lower body temperature, leading to death from hypothermia. Hypothermia is a medical emergency that requires immediate treatment otherwise it can quickly kill people. Winds and heavy wet snow or icing may bring down trees and power lines, leaving affected residents without heat, which can result in death from hypothermia also. People also die from falling ice itself, which can weigh as much as a large rock and hit the ground with enough velocity to kill instantly. This is a major problem around skyscrapers where officials may have to close entire city blocks to foot and vehicle traffic due to ice falling from great heights. People also kill by roof and building collapse from the weight of snow. Large amounts of snow and ice on a building's roof can strain the structure to fail. In such a situation, commercial buildings with large, flat roofs are especially liable to collapse and ultimately cause deaths. Finally, winter storm deaths do occur from carbon monoxide poisoning and fires caused by improper use of fireplaces and other heat sources (Bourque et al. 2007).

Volcanoes

Some volcanic eruptions have killed thousands of people, while others have killed few people. For example, the 1815 Tambora volcanic eruption in Indonesia killed more than 92,000 people. Even as recently as 1985, the volcano Nevado del Ruiz ravaged the Colombian town of Armero, killing nearly 25,000 people. However, often because of remote location, large volcanic eruptions sometimes do not kill people. For example, no one died from the 1912 Alaska eruption because no one lived nearby. Overall, compared to other natural disasters, the number of human fatalities is relatively quite low. Ebert (2000) noted that over the past 500 years, volcano-related disasters caused about 200,000 fatalities, which, on average, translates to 400 deaths per year. Supporting this relatively low number, a recent study from the University of Bristol reported that more than 278,000 died from volcanic activities from 1500 to 2017, amounting to 540 deaths per year on average (Fosco 2017).

Volcanoes provide advance warning signs of imminent eruptions which partially explains the relatively low fatalities. After observing these signs, people either on their own leave the vicinity or are instructed by the government to leave. Also, it is easy to leave safely during an eruption if it involves mainly lava flows (Ebert 2000).

The University of Bristol study further analyzed the distance each person was from the volcano when they died and causes of death. The study found that nearly 50 percent of all volcanic deaths occurred within 7 miles (10.5 km) of the eruption, but fatalities also occurred as far away as 106 miles (159 km). The most common cause of death was volcanic bombs or ballistics for those within 3 miles (4.5 km). "Between 3–10 miles, an avalanche of hot rock, ash, and gas, known as pyroclastic flow, was the most common cause of death, while volcanic mudslides, tsunamis, and ash fall are the main danger at greater distances" (Fosco 2017; also see Salleh 2001).

Referring to Professor Russell Blong's work, Salleh (2001) reported that since the year 1500, 274,000 deaths were caused by 400 volcanic eruptions. Twenty-eight percent of these were caused by pyroclastic flow, which can travel up to 133 miles (200 km) an hour. People also die due to starvation as were the cases in the 1815 Tambora, Indonesia earthquake, and the 1783 Laki, Iceland earthquake. Post-eruption disease outbreak also causes deaths as was the case in the 1991 earthquake in Pinatubo, Philippines. According to Blong, these two indirect causes were the second most common causes of death for the above 400 volcanic eruptions, accounting for 23 percent of all deaths. The third most common cause of death (20 percent) was from tsunamis triggered by volcanic material falling into the sea. For example, 90 percent of the deaths from the 1883 Krakatoa eruption resulted from a tsunami. Additionally, mud flows killed 13 percent of people, and another 3 percent were killed by falling ash that spread over a large area and caused

Table 4.3 Major causes of deaths for selected volcanic eruptions, AD 79–1991

Frequency	*Average number of death*	*Cause*
2	50,675	Starvation
3	12,860	Tsunami
8	5,204	Ash fall
6	6,170	Mudflow
5	4,627	Ash flow, Ash fall; Mudflow, Tsunami; Lava flow, Roof collapse and disease*

*Combination of two or three causes

Sources: Complied from Blong (1984, 424) and other sources.

the collapse of buildings. Seven to eight percent of deaths had an unknown cause, and less than 1 percent of total deaths were caused by lava (Salleh 2001). Although permanent residents were among the overwhelming number of deaths caused by volcanic activities, tourists, scientists (mostly volcano experts), emergency responders, and media personnel have also died. Table 4.3 lists the major causes of death-induced by volcanic eruptions. The table includes 24 such events. Based on average deaths, the most important cause of volcano related death is starvation, followed by tsunami, ash flows, and mudflows. Deaths were also caused by any of a combination of two causes: ash flows and falls; ash flows and mudflows; volcano collapse and tsunami; mudflows and lava flows; and roof collapse and disease. To be specific, causes of death for the May 18, 1980, eruption of Mt. St. Helens in the United States are provided here. This was the most destructive volcanic eruption in the country, killing 57 people and thousands of animals. Direct causes of death included asphyxiation by dense ash exposure, burns, falling from trees, and trauma. Four indirect deaths were caused by a crop duster hitting powerlines during the ash fall, two heart attacks from shoveling ash, and a traffic accident during poor visibility (Carson, 1990; Simkin and Siebert 1994).

Other natural disasters

Other natural disasters such as avalanches, landslides, hail, and wildfires kill a relatively small number of people annually. For example, avalanches killed 22 people per year in the world between 2006 and 2015 (IFRC 2016). In the United States, 925 avalanche fatalities were reported between 1951 and 2013 with an average of 15 ± 11 per year (mean ± SD; range, 0–40 fatalities per year) (Jekich et al. 2016). Highest deaths recorded in Colorado, which accounted for 27 percent of avalanche fatalities. Other states experienced deaths were: Alaska, Montana, Utah, Washington, and Wyoming. In the United States avalanche fatalities are most common during December to April months. However, fatalities can occur in every month of the year. This is also true for Europe.

Avalanches are a type of mass movement with two major subtypes: rock avalanches and snow avalanches. Both types of mass movement occur mostly in inaccessible hilly and mountainous areas. Deaths result from the former type of avalanches in several major ways: blunt force trauma from collisions with objects, being hit by falling debris, or suffocation. In the case of snow avalanches, asphyxiation (or asphyxia) is the primary cause of death. Asphyxiation is a condition of deficient supply of oxygen to the body that arises from abnormal breathing. McIntosh et al. (2007) examined the causes of death in avalanche fatalities in Utah from the 1989–90 to 2005–06 winter seasons. They reported that most deaths occurred during recreational backcountry activities: 85.7 percent of deaths were due to asphyxiation, 8.9 percent were due to a combination of asphyxiation and trauma, and 5.4 percent were due to trauma alone. Head injuries were frequent in those killed solely by trauma (also see Sheet et al. 2018).

Landslides killed 9,477 people in the decade from 2006 to 2015, averaging nearly 948 people per year (IFRC 2016). Most landslide fatalities are from rock falls, debris-flow, or volcanic debris flows, called lahars. Landslides are triggered by earthquakes, volcanic eruptions, heavy rainfall on steep hillsides or by wildfires. The USGS claims that on average, 25–50 people are killed by landslides each year in the United States. Both in cases of avalanches and landslides, the causes of death are similar, and for both disasters, fatalities are underestimated because many people who die from avalanche- and landslide-caused injuries do so long after the event itself (Petley 2012).

Hail has killed people around the world throughout history. Sometimes the death toll has reached into the hundreds, but those incidents are rare now. For example, in the United States, only four people have been killed by hail since 2000. Two of these deaths happened in 2000, another in 2005, and one more in 2008. Before 2000, on average, only one death occurred per year in the USA due to hail. Globally, the deadliest most recent incident of hail-associated deaths was reported in the state of Andhra Pradesh, India in 2013 where at least nine people were killed throughout several villages (Shaw 2016). Hail causes more injuries than deaths, and whether one or the other depends on the size of the hail, the wind speed, and the frequency with which the stones fall. Specifically, deaths often occur from strikes to the head or other vulnerable parts of the body.

Not all countries of the world are prone to wildfires; however, the United States and Australia are highly vulnerable, especially in relatively unpopulated areas. In the USA, on average, approximately 47 people die from wildfires per year. Most deaths are caused by smoke inhalation and burns (Bourque et al. 2007). As of January 8, 2020, the 2019–2020 Australia's devastating wildfire season (September to January) killed 28 people. In addition, over one billion animals (e.g., koalas, kangaroos, wallabies, red foxes, and feral cats) had been wiped out. The figure includes mammals (excluding bats), birds and reptiles. It does not include frogs, insects or other invertebrates (Samuel 2020).

Conclusion

Analysis of causes and circumstances of disaster-induced human deaths suggests that a large proportion of these fatalities are largely preventable. For example, most of the drowning deaths from storm surges, floods, and tsunamis are avoidable by improving cyclone, tsunami, and flood forecasting and warning systems, and evacuation and sheltering measures. In fact, many disaster experts stress that one of the effective ways to escape a massive storm surge is to evacuate people from the coastline or rivers. Construction of concrete seawalls and levees along the coast can also reduce loss of life and property from storm surges, tsunamis, and cyclones/hurricanes/typhoons. In developed countries, over half of drowning deaths from these disasters occur due to driving automobiles into flood water. Partly, this is because people are not always aware of the power of storm surges and flash floods. Rushing water can float many large objects, including cars. The debris in the rushing water can also cause deaths. Thus, proper education for at-risk population can decrease drowning deaths from all hydrological/meteorological disasters.

Adequate preparation for any natural disaster can also minimize human deaths. For example, when a blizzard warning is provided, people should stock up on non-perishable foods, blankets, flashlights, extra batteries, and candles beforehand, strictly avoid travel, and stay indoors. If travel is necessary by car during a blizzard, it is vital to have an emergency aid kit (water, jumper cables, road flares, tow rope, warms clothes, and non-perishable snacks) in case the car breaks down, gets into an accident, or becomes stuck in snow. To avoid hypothermia if caught outdoors during a blizzard, people should stay hydrated and nourished. People also need to keep blood flowing by moving around.

For some natural disaster like avalanches, and landslides rescue strategies that employ rapid recovery as well as techniques that prolong survival while buried are the best means to improve survival outcome. In the case of an earthquake, some people lost in the rubble can be saved if effective search and rescue operations quickly reach destroyed and damaged buildings, and infrastructure. Also, deaths from earthquakes can be reduced significantly if construction of buildings follows earthquake resistant technology.

Notes

1. Several hospitals records found that the injury-to-death ratio varies among earthquake disasters but averages about three injuries to every fatality (NHO 2012).
2. Thrust earthquakes occur where tectonic plates move vertically up and down and displace ocean water, while strike-slip earthquakes occur where tectonic plates move horizontally.
3. In the 1992 Puerto Rico floods, 11 of the 14 fatal car crashes occurred when drivers were crossing flooded bridges (Staes et al. 1994).

4. There are other disaster events where rescuers died during rescue operations. For example, five rescuers died in North Carolina when Hurricane Floyd hit in 1999 (MMWR 2000).
5. Compiled from different sources, average annual flood mortality in Bangladesh was 205 persons during the 1972–2019 period.
6. It was the largest cause of death when Hurricane Katrina hit in 2005. Most deaths occurred because of drowning.
7. During the last 20-year period (1998–2018), a total of 2,255 people died in the United States from hurricanes, averaging near 113 deaths per year (www.iii.org/fact-statistic/facts-statistics-hurricanes).
8. With few exceptions, relatively more indirect deaths occur from cyclones/hurricanes/typhoons in developing countries than in developed countries.
9. Indoor deaths in the continent of North America showed a higher percentage in comparison with Hurricane Katrina and Rita (Jonkman et al. 2009).
10. This represents 5 percent of the total deaths caused by Cyclone Gorky.
11. One police officer was also killed by lightening while he was assisting with recovery and cleanup efforts the day after the tornado. This indirect death was not included in the official death toll.
12. A district is the second largest administrative division in Bangladesh with an average population of 2.5 million.
13. A proximate death is a direct death, while the underlying cause of death refers to the disease or injury that ultimately leads to a death.

References

Ahern, M., R.S. Kovats, R. Few, and F. Matthies. 2005. Global Health Impacts of Floods. *Epidemiologic Reviews* 27(1): 36–46.

Alirol, E., S.K. Sharma, H.S. Bawaskar, U., Kutch, and F. Chappuis. 2010. Snake Bite in South Asia: A Review. *PLoS Neglected Tropical Diseases* 4(1): e603.

Amadeo, K. 2019. Heat Waves and Their Effect on the Economy: How Much Do Heat Waves Cost Us? How Much Worse Will They Get? (www.thebalance.com/heat-wave-causes-list-effect-on-the-economy-4173881 – last accessed November 14, 2019).

Below, R., E. Grover-Kopec, and M. Dilley. 2007. Documenting Drought-Related Disasters: A Global Reassessment. *The Journal of Environment Development* 16(3): 328–344.

Blong, R.J. 1984. *Volcanic Hazards: A Sourcebook on the Effects of Eruptions*. Orlando, FL: Academic Press.

Bolt, B.H. 2004. *Earthquakes*. New York, NY: W.H. Freeman and Company.

Bourque, L.B., J.M. Siegel, M. Kano, and M.M. Wood. 2007. Morbidity and Mortality Associated with Disasters. In *Handbook of Disaster Research*, edited by H. Rodriguez, E.L. Quaranteli, and R.R. Dynes, pp. 97–112. New York, NY: Springer.

Brown, S., P. Archer, E. Kruger, and S. Mallonee. 2002. Tornado-Related Deaths and Injuries in Oklahoma due to the 3 May 1999 Tornadoes. *Weather and Forecasting* 17(3): 343–353.

Cardona, D.O. 2004. The Need for Rethinking the Concept of Vulnerability and Risk from a Holistic Perspective: A Necessary Review and Criticisms for Effective Risk Management. In *Mapping Vulnerability: Disasters, Development and People*, edited by Bankoff, G., G. Frerks, and D. Hilhorst, pp. 37–51. London: Earthscan Publishers.

Carson, R. 1990. *Mount St. Helens: The Eruption of Recovery of a Volcano*. Seattle, WA: Sasquatch Books.

CDC (Centers for Disease Control and Prevention). 2012. Tornado-related Fatalities – Five States, Southeastern United States, April 25–28, 2011. *Morbidity and Mortality Weekly* 28: 529.

CDC (Centers for Disease Control and Prevention). 2013. Deaths Associated with Hurricane Sandy – October–November 2012. *Morbidity and Mortality Weekly Report (MMWR)* 62(20): 393–397.

CDC (Centers for Disease Control and Prevention). 2017. *Death Scene Investigation After Natural Disaster or Other Weather-Related Events Toolkit*. Atlanta.

CRED (Centre for Research on the Epidemiology of Disasters) and UNISDR (United Nations International Strategy for Disaster Reduction). 2016. *The Human Cost of Weather Related Disasters: 1995–2015*. Brussels: CRED.

CRED (Centre for Research on the Epidemiology of Disasters), USAID (United States Aid for International Development), and UNISDR (United Nations International Strategy for Disaster Reduction). 2016. *The Human Cost of Natural Disasters 2015: A Global Perspective*. Brussels: CRED.

Chang, K-C., and C-T. Chang. 2019. Using Cluster Analysis to Explore Mortality Patterns Associated with Tropical Cyclones. *Disaster* 43(4): 891–905.

Chiu, C.H., A.H. Schnall, C.E. Mertzlufft, R.S. Noe, A.F. Wolkin, J. Spears, M. Casey-Lockyer, and S.J. Vagi. 2013. Mortality From a Tornado Outbreak, Alabama, April 27, 2011. *American Journal of Public Health* 103(8): e52–e58.

Coburn, A.W., R.J.S. Spence, and A. Pomonis. 1992. Factors Determining Human Casualty Levels in Earthquakes: Mortality Prediction in Building Collapse. *Earthquake Engineering Tenth World Conference*, 5989–5994.

Combs, D.L., R.G. Parrish, S.J.N. McNabb, and J.H. Davis. 1996. Deaths Related to Hurricane Andrew in Florida and Louisiana. *International Journal of Epidemiology* 25(3): 537–544. Rotterdam.

Cross, R. 2015. Nepal Earthquake: A Disaster that Shows Quakes Don't Kill People, Buildings Do. The Guardian, 13 May (www.theguardian.com/cities/2015/apr/30/nepal-earthquake-disaster-building-collapse-resilience-kathmandu – last accessed November 13, 2019).

CRS (Congressional Research Service). 2014. *Typhoon Haiyan (Yolanda): U.S. and International Response to Philippines Disaster*. Washington, DC.

Daniel, M. 2012. Who Died During Hurricane Sandy, and Why? November 27 (https://earthsky.org/earth/who-died-during-hurricane-sandy-and-why – last accessed November 21, 2019).

Diakakis, M., G. Deligiannakis, K. Katsetsiadou, and E. Lekkas. 2015. Hurricane Sandy Mortality in the Caribbean and Continental North America. *Disaster Prevention and Management* 24(1): 132–148.

DKKV (German Committee for Disaster Reduction) (ed.). 2012. *Detecting Disaster Root Causes: A Framework and an Major Analytical Tool for Practitioners*. Bonn: DKKV Publication Series, 48.

Ebert, C.H.V. 2000. *Disasters: Violence of Nature, Threats by Man*. Dubuque: Kendall/Hunt Publishing Company.

FitzGerald, G. Du, W. Jamal, A., M. Clark, M, and X. Hou. 2010. Flood Fatalities in Contemporary Australia (1997–2008). *Emergency Medicine Australasia* 22(5): 180–186.

Fosco, M. 2017. 500 Years of Volcano Deaths Could Help Save the 800M People at Risk Today, 10 June (www.seeker.com/earth/500-years-of-volcano-casualty-data-could-help-improve-safety-around-eruptions – last accessed February 9, 2019).

Gupta, K. 2009. Cross-Cultural Analysis of Response to Mass Fatalities following 2009 Cyclone Aila in Bangladesh and India. *Quick Response Report #216. Natural Hazards Center*, University of Colorado at Boulder, CA.

Haque, C.E., and D. Blair. 1992. Vulnerability to Tropical Cyclones: Evidence from the April 1991 Cyclone in Coastal Bangladesh. *Disasters* 16(3): 217–229.

Horstmann, B. 2004. *Glacial Lake Outburst Floods in Nepal and Switzerland: New Threat due to Climate Change*. Bonn: Germanwatch.

Holle, R.L. 2016. A Summary of Recent National-Scale Lightning Fatalities Studies. American Meteorological Society (https://doi.org/10.1175/WCAS-D-15-0032.1 – last accessed November 23, 2019).

IFRC (International Federation of Red Cross and Red Crescent Societies). 2016. *2016 World Disasters Report: Resilience: Saving Lives Today, Investing for Tomorrow*. Geneva: International Federation of Red Cross and Red Crescent Societies.

Jekich, B.M., B.D. Drake, J.Y. Nacht, A. Nichols, A.A. Ginde, and C.B. Davis. 2016. Avalanche Fatalities in the United States: A Change in Demographics. *Wilderness & Environmental Medicine* 27(1): 46–52.

Jonkman, S.N., and I. Kelman. 2005. An Analysis of the Causes and Circumstances of Flood Disaster Deaths. *Disasters* 29(1): 75–97.

Jonkman, S.N., B. Maaskant, E. Boyd, and M.L. Levitan. 2009. Loss of Life Caused by the Flooding in New Orleans after Hurricane Katrina: Analysis of the Relationship between Flood Characteristics and Mortality. *Risk Analysis* 29(5): 676–698.

Kuligowski, E.D. 2020. Field Research to Application: A Study of Human Response to the 2011, Joplin Tornado and Its Impact on Alerts and Warning in the USA. *Natural Hazards* 102: 1057–1076.

Kuligowski, E.D., F.T. Lombardo, L.T. Phan, M.L. Levitan, and D.P. Jorgensen. 2013. *Technical Investigation of the May 22, 2011, Tornado in Joplin, Missouri: Draft Final Report, National Institute of Standards and Technology (NIST)*. Washington, DC: NIST.

Kunii, O., S. Nakamura, R. Abdur, and S. Wakai. 2002. The Impact on Health and Risk Factors of the Diarrhoea Epidemics in the 1998 Bangladesh Floods. *Public Health* 16(2): 68–74.

McIntosh, S.E., C.K. Grissom, C.R. Olivares, H.S. Kim, and B. Tremper. 2007. Cause of Death in Avalanche Fatalities. *Wilderness Environmental Medicine* 18(4): 293–297.

McPherson, M., M.M. Counahanb, and J.L. Hallb. 2015. Responding to Typhoon Haiyan in the Philippines. *Western Pacific Surveillance and Response Journal* 6(S 1): 1–4.

Mersereau, D. 2013. Why So Many People Died from Haiyan and Past Southeast Asia Typhoon, 11 November. *The Washington Post* (www.washingtonpost.com/news/capital-weather-gang/wp/2013/11/11/inside-the-taggering-death-toll-from-haiyan-and-other-southeast-asia-typhoons/ – last accessed August 6, 2019).

Mersereau, D. 2016. 7 Common Causes of Death During Winter Storms (www.mentalfloss.com/article/73513/7-common-causes-death-during-winter-storms-and-how-prevent-them – last accessed November 15, 2019).

MMWR (Morbidity and Mortality Weekly Report). 2000. Morbidity and Mortality Associated With Hurricane Floyd – North Carolina 49(17): 369–372.

Mulder, M.B., L. Msalu, T. Caro, and J. Salerno. 2012. Remarkable Rates of Lightning Strike Mortality in Malawi. *PLOS ONE* 7(1): e2981.

NHO (Natural Hazards Observer). 2012. Crush Injuries Kill in Earthquakes. March, 5.

Nishikiori, N., T. Aba, D.G.M. Costa, S.D. Dharmaratne, O. Kunii, and K. Moji. 2006. Who Died as a Result of the Tsunami? Risk Factors of Mortality Among Internally Displaced Persons in Sri Lanka: A Retrospective Cohort Analysis. *BMC Public Health* 6: 73. https://doi.org/10.1186/1471-2458-6-73.

Noji, E.K. 1997. Earthquake. In *Public Health Consequences of Disasters*, edited by Noji, E.K. Oxford: Oxford University Press.

Padli J, M.S. Habibullah, and A.H. Baharom. 2013. Determinants of Flood Fatalities: Evidence from a Panel Data of 79 Countries. *Social Science & Humanities* 21: 81–98.

Paul, B.K. 1998a. Coping Mechanisms Practised by Drought Victims (1994–95) in North Bengal, Bangladesh. *Applied Geography* 18(4): 355–373.

Paul, B.K. 1998b. Coping with the 1996 Tornado in Tangail, Bangladesh: An Analysis of Field Data. *The Professional Geographer* 50(3): 287–301.

Paul, B.K. 2002. Flash Flooding in Kansas: Respondents Satisfaction with Emergency Response Measures and Disaster Aid. *Great Plains Research* 12(2): 295–322.

Paul, B.K., and S. Dutt. 2010. Hazard Warnings and Responses to Evacuation Orders: The Case of Bangladesh's Cyclone Sidr. *Geographical Review* 100(3): 336–355.

Paul, B.K., and M. Stimers. 2012. Exploring Probable Reasons for Record Fatalities: The Case of 2011 Joplin, Missouri, Tornado. *Natural Hazards* 64(2): 1511–1526.

Paul, B.K., and Stimers, M. 2014. Spatial Analyses of the 2011 Joplin Tornado Mortality: Deaths by Interpolated Damage Zones and Location of Victims. *Weather, Climate and Society* 6(2): 161–174.

Petley, D. 2012. Global Patterns of Loss of Life from Landslides. *Geology* 40(10): 927–930.

Rappaport, E.D. 2014. Fatalities in the United States from Atlantic Tropical Cyclone: New Data and Interpretation. *Insights and Innovations*, March: 341–346.

Rappaport, E.D., and B.W. Blanchard. 2016. Fatalities in the United States Indirectly Associated with Atlantic Tropical Cyclones. *Insights and Innovations*, July: 1139–1148.

Salerno, J., L. Msalu, T. Caro, and M.B. Mulder. 2012. Risk of Injury and Death from Lightning in Northern Malawi. *Natural Hazards* 62(3): 853–862.

Salleh, A. 2001. How Volcanoes Kill, 17 January (www.abc.net.au/science/articles/2001/01/17/234135.htm – last accessed November 14, 2019).

Samuel, S. 2020. A Staggering 1 Billion Animals are Now Estimated Dead in Australia's Fires, January 7 (www.vox.com/future-perfect/2020/1/6/21051897/australia-fires-billion-animals-dead-estimate – last accessed January 8, 2020).

Saulnier, D.D., H.K. Green, T.D. Waite, R. Ismail, N.B. Mohamed, C. Chhorvann, and V. Murray. 2019. *Disaster Risk Reduction: Why Do We Need Accurate Disaster Mortality Data To Strengthen Policy And Practice?* New York, NY: UN Office for Disaster Risk Reduction.

Schencking, C. 2013. *The Great Kanto Earthquake and the Chimera of National Reconstruction in Japan*. New York, NY: Columbia University Press.
Seidl, S. 2006. Pathological Features of Death From Lightning Strike. *Forensic Pathology Reviews* 4: 3–23.
Shaw, J. 2016. 8 Deadliest Hail Storms in History. Newsmax, 8 July (www.newsmax.com/FastFeatures/hail-deadliest-storms/2016/07/08/id/737837/ – last accessed November 25, 2019).
Sheet, A., D. Wang, S. Logan, and D. Atkins. 2018. Causes of Deaths Among Avalanche Fatalities in Colorado: A 21-Year Review. *Wilderness Environmental Medicine* 29(3): 325–329.
Shultz, J.M., J. Russell, and Z. Espinel. 2005. Epidemiology of Tropical Cyclones: The Dynamics of Disaster, Disease, and Development. *Epidemiologic Review* 27(1): 21–35.
Simkin, T., and L. Siebert. 1994. *Volcanoes of the World*. Tucson: Geoscience Press.
Singh, O., P. Bhardwaj, and J. Singh. 2017. Distribution of Lightning Casualties over Maharashtra, India. *Journal of Indian Geophysical Union* 21(5): 415–424.
Staes, C., J.C. Orengo, J. Malilay, J. Rullan, and E. Noji. 1994. Deaths Due to Flash Floods in Puerto Rico, January 1992: Implications for Prevention. *International Journal of Epidemiology* 23(5): 968–975.
Statista. 2019. Natural Disasters in the U.S. – Statistics & Facts (www.statista.com/topics/1714/natural-disasters/ – last accessed November 14, 2019).
Sugimoto, J.B., A.B. Labrique, S. Ahmad, M. Rashid, A.A. Shamim, B. Ullah, R.D.W. Kleman, P. Chistian, and K.P. West, Jr. 2011. Epidemiology of Tornado Destruction in Rural Northern Bangladesh: Risk Factors for Death and Injury. *Disasters* 35(2): 329–345.
Szczepanski, K. 2017. The Great Kanto Earthquake in Japan, 1923 (www.thoughtco.com/the-great-kanto-earthquake-195143 – last accessed, November 13, 2019).
Telford, J., J. Crsgrave, and R. Houghton. 2006. *Joint Evaluation of the International Response to the Indian Ocean Tsunami: Synthesis Report*. London: TEC.
Williams, J. 2013. How Lightning Kills and Injuries Victims. *The Washington Post*, June 27 (www.washingtonpost.com/news/capital-weather-gang/wp/2013/06/27/how-lightning-kills-and-injures-victims/ – last accessed November 23, 2019).
WCD (World Congress on Drowning). 2003. Recommendations (www.drowning.nl. – last accessed November 11, 2019).

5 Determinants of disaster deaths

This chapter covers a discussion of risk factors associated with deaths caused by various natural disasters. The study of such factors provides valuable insights about areas where outreach, mitigation, and prevention can be done to help reduce the loss of life. The discussion in this chapter generally follows the order of extreme events presented in Chapter 3. For convenience, storms are disaggregated into tropical cyclones/hurricanes/typhoons, tornadoes, blizzards, and lightning, and under extreme temperatures, determinants of deaths resulting from heat waves and wildfires are presented.

Generally, the factors or determinants of deaths are studied by using bivariate (e.g., chi-square, *t*-test, and, correlation) and/or multivariate (e.g., multiple regression and its variants) techniques. Ideally, multivariate techniques are appropriate to identify relative influence of one relevant risk factor compared to other factors. That is, the effect of a particular independent variable is made more certain, for the possibility of distorting influences from the other independent variables is removed. In fact, determinants of disaster deaths are complex because they interact with each other. In both bivariate and multivariate analyses, often odds ratios (ORs) are calculated. The OR is a useful measure of association between two dichotomous variables – one is called an exposure (e.g., smoking) and the other is called an outcome (e.g., lung cancer).[1] The OR is calculated by dividing the odds of the first group by the odds in the second group (Card 2012). Occasionally, OR will be mentioned in context of risk of deaths.

Toward a general framework

To present the determinants of deaths caused by the major types of natural disasters, a general framework is in order. In this case, building one requires first an overview of the classic epidemiologic model of agent, host, and environment. This model is known as the epidemiologic triad or the epidemiologic triangle (Figure 5.1). It was originally developed to identify causes of infectious diseases (Smith 1934), and a considerable number of disaster researchers (e.g., Parrish et al. 1964; Logue et al. 1981; Ramirez and Peek-Asa 2005; Paul and Ramekar 2018) have applied this model to select and examine the determinants of earthquake-induced mortality and injury. The three apexes of the triangle of this model are the host (who), the agent (what), and the environment (where). In the original model, host is either humans or animals that are exposed to and harbor a disease. The agent is the concerned disease, and the environment is the favorable surroundings and conditions external to the host that cause or allow the disease to be transmitted (Smith 1934).

Variants of the epidemiologic triad have been used by a number of health/medical geographers (e.g., Meade 1977; Keil and Ali 2006; Donaldson and Wood 2008; Carrel and Emch 2013; Blackburn et al. 2014; Emch et al. 2017) in the context of the state of human health, and the emergence and reemergence of human pathogens across space through time. Their works rightly belong to the "disease ecology" or "geographic pathology" approach to traditional medical geography. Modifying this approach to the context of earthquake mortality and morbidity, hazard researchers (e.g., Armenian et al. 1997; Peek-Asa et al. 2003; Ramirez and Peek-Asa 2005; Paul and Ramekar 2018) have included demographic and socioeconomic conditions, behavior, and physical location as host factors.

It is worthwhile to mention that the study unit determines who is host, and the host, in turn, determines the selection of appropriate variables. Host could be an individual, a household, or a geographic area at any one of several scales (e.g., county, state, region, and country). For example, yearly annual income could be a host factor if the study unit is either individual or household. If the study unit is a geographic area other than a country, the appropriate income variable is median or mean yearly income of the selected geographic unit. When a country is the study unit, Gross Domestic Product (GDP) per capita can be used to represent national income.

Behavior, on the other hand, refers only to individual- and household-level studies, and this factor includes what individuals or proportion of household members do during the earthquake: Do they stay inside the house? Move to open space? Turn off oil stoves to prevent fires? Protect property from falling? Meanwhile, physical location as a host factor refers to distance from the epicenter to the location of individuals or individual households. When the study unit is geographical, distance is generally measured from the epicenter to the geographical center of the study unit. Depending on the areal scale of the study area, the extent of mortality differs by other host factors such as rural versus urban residence and developed versus developing countries.

In the case of natural disasters, agent factors include physical characteristics of each type of extreme event (e.g., magnitude, duration, frequency, temporal spacing, spatial extent, spatial dispersion, speed of onset, and diurnal factors),[2] while environmental factors include relevant characteristics of both built and physical environments within which people live (Figure 5.1). The former includes such characteristics as building materials of residential and nonresidential structures, age, height (number of floors),

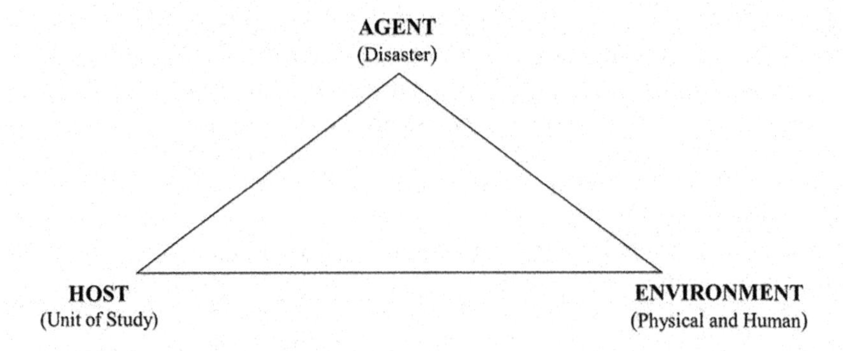

Figure 5.1 Epidemiologic triangle.

Source: Based on Smith (1934).

and size of buildings. In the context of earthquakes, all these build factors are associated with deaths caused by such events (Ramirez and Peek-Asa 2005). Physical environment-related factors include soil types and topographic conditions of the disaster-affected areas. For example, an area with alluvial soil, which has high water content, may exhibit greater damage, injuries, and deaths from earthquakes than an area with dry and compact soil (Ramirez and Peek-Asa 2005).

For general discussion on determinants of different types of disaster, the epidemiological triad model is integrated with hazard and vulnerability factors originally proposed by Jonkman and Kelman (2005) to analyze the causes and circumstances of flood deaths in Europe and the United States. Later, the model was modified by Paul (2011) and Paul and Mahmood (2016). As shown in Figure 5.2, determinants of disaster-induced deaths are

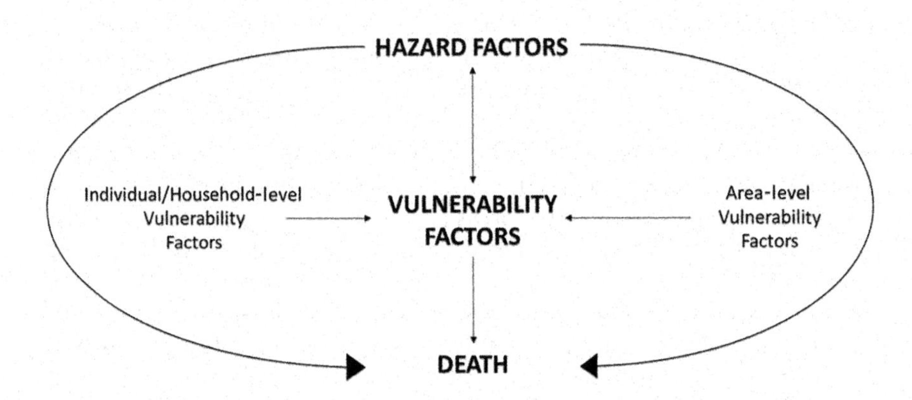

Figure 5.2 Hazard and vulnerability factors as determinants of disaster-induced deaths.

Sources: Modified after Paul (2011), and Paul and Mahmood (2016).

broadly classified into two: Hazard and vulnerability factors. The former includes physical characteristics/parameters of the extreme event under consideration. On the basis of study unit, the vulnerability factors are also divided into two groups: Individual/household factors and factors related to areal study unit. Family income, literacy, age, gender, disability, and house type are some examples of individual/household-level vulnerability factors. These factors include both demographic and socioeconomic conditions as well as built environment. When areal unit is the study unit, vulnerability factors include median income/GDP per capita, level of poverty, urbanization rate, population density, land area, population size, level of development, and political system. Physical environment is also part of this study unit. However, the broad factors are neither simple in themselves nor should they stand alone. Also, these two factors often simultaneously affect disaster deaths, and thus, they are connected to each other with arrows (Figure 5.2).

Earthquakes

Like all other disaster deaths, determinants of earthquake mortality rates depend on the type and unit of study. Type here means whether one is interested in study of a particular earthquake in a particular year, a considerable number of earthquakes in a given year, earthquakes in several years, or any combination of the above. Unit of study varies from individual/household levels to different areal units such as county, district, state, or national levels. Earthquake deaths are a function of physical dimension of earthquake or agent factor: Magnitude, intensity, duration, scale (spatial extent), temporal spacing (e.g., time of strike, day of the week, and time of year), epicentral distance, that is, distance from the epicenter to the study unit, and secondary hazards (e.g., aftershocks, landslides, and tsunamis). Moreover, magnitude and intensity of earthquake are measures of an earthquake's strength (Paul 2011). The former is generally measured by using the Richter scale and the latter by using peak ground acceleration or Modified Mercalli Intensity (MMI), which provides a useful relative indication of an earthquake based on subjective human experience (Paul 2011). Additionally, peak ground acceleration, ground motion, or shaking can be quantified as seismic energy based on the speed of acceleration of the ground at localized points (Ramirez and Peek-Asa 2005).

Meanwhile, magnitude and intensity measure, and duration of earthquake are of limited value for studies of a single earthquake because they provide only one number for the entire event. Similarly, time of day and day of the week should not be included in the context of single earthquake studies. However, magnitude, intensity, duration, and area affected by an earthquake are directly related to deaths, while time of day/strike affects earthquake mortality rates. For example, if an earthquake occurs at night in an area where buildings can collapse easily, this can result in a higher death toll because most people are in bed. Across the globe, the primary cause of death from earthquakes is falling debris caused by structural collapse of buildings and infrastructure (Noji et al. 1990; Petal 2011; Shapira et al. 2015). However, while the 1994 Northridge, California earthquake occurred at night, it killed only 57 people because of its strict building codes that decreased building collapse (Mathue-Giangreco et al. 2001). Had this earthquake occurred during the day, more "people could have been exposed to different environmental hazards in the workplace, at schools, in public locations, or on roadways or thoroughfares" (Ramirez and Peek-Asa 2005, 350). As a result, deaths occurring because of earthquake would likely have increased because many more people would have been using transportation systems or been located in large, multiple-story buildings.

The day of the week also affects earthquake mortality. While analyzing field data collected from Nepal, Paul and Ramekar (2018) noted that had the April 25, 2015, earthquake occurred on a weekday, both deaths and injuries, particularly among children, would likely have been higher. This earthquake occurred at noon on a weekly holiday when many Nepalese, including children, were outdoors. In contrast, 19,000 of the 78,000 deaths caused by the 2005 Kashmir earthquake were children who were attending schools. Most of these children died because of widespread collapses of school buildings. The earthquake took place shortly after 8:50 am when children were in school (Paul 2011). If the earthquake had taken place after school hours or during a weekend, fewer children might have died. Being indoors or outdoors during earthquakes, and time or day of occurrence also can affect the number of resulting fatalities because most deaths are the result of damage to buildings or other structures. For instance, during the 1988 Armenian earthquake, deaths were substantially elevated among those who remained indoors than those who exited buildings (Armenian et al. 1992).

The time of year or temporal spacing is also a determinant of earthquake-induced mortality in part because seasonal differences in temperature can exacerbate the health effects of an earthquake. For instance, during winter or rainy seasons, mortality is generally high for people whose houses are either damaged or destroyed by earthquakes. Additionally, bad weather also hampers rescue efforts and makes necessary recovery much harder. However, an earthquake occurring in warm dry weather also presents several specific problems. Warm, dry air can facilitate more rapid decay of bodies, and this, in turn, leads to an increase in the spread of diseases following the earthquake, particularly in areas where clean water is scarce. Similar to magnitude, intensity, and duration, temporal spacing should not be included in the context of single earthquake studies.

Although increasing distance from the epicenter is inversely related with earthquake mortality rates, Peek-Asa et al. (2000) found that distance was a poor predictor of death in the case of the 1994 Northridge earthquake. This could be explained by the localized strength of shaking, which is influenced by ground materials, liquefaction, and landslide susceptibility

(Alexander 1993; Seligson and Shoaf 2003). Generally, earthquake energy waves transmit more effectively through alluvial soil with high moisture contentment than through dry soil and areas of landfill. Thus, the mortality is likely to be higher in areas where soil has a high water content than the areas with dry and compact soil. Sandy and soggy soil can liquefy if shaking is strong and lasts long enough.[3] Two studies (Mathue-Giangreco et al. 2001; Peek-Asa et al. 2003) found a positive, but statistically nonsignificant association between rock structures and casualties during the 1994 Northridge earthquake. Soil and rock types, and topographic conditions are all related to (physical) environmental factors.

Major earthquakes are usually followed by many aftershocks, which have potential to cause further damage, especially to structures weakened by the initial earthquake. Moreover, if the topography of the affected region, particularly around the epicenter, is rugged and/or contained steep slopes, such topography and slope increase the chances for aftershock landslides, which, in turn, pose threat to structures and people of the region (Margesson and Taft-Morales 2010). In contrast to aftershocks, foreshocks provide early warnings of the major earthquake. For example, within 24 hours prior to the main shock, a series of three foreshocks hit Chile on May 22, 1960 that provided early warnings of the imminent major earthquake (Brumbaugh 1999). However, aftershocks play an important role in saving lives also because the main event sends many people from their homes to the street or open space. Often, they spend most of their time outdoor, and sleep for several days in tents.

Vulnerability or host factors are referred to as individual/household demographic and socio-economic characteristics, or characteristics of selected areal units, ranging from communities to countries. Many studies (e.g., Khan and Mustafa 2007; Green and Miles 2011; Johnston et al. 2014; Shapira et al. 2015; Lambie et al. 2016) have reported that persons at elevated risk of death include females, children and the elderly, and people with disabilities. For example, most victims in the Great Hanshin or Kobe earthquake were elderly (Masai et al. 2009). Meanwhile, in the 1993 earthquake in India, females comprised 55 percent of fatalities but made up only 48 percent of the population, and female deaths outnumbered male deaths in all age cohorts (Krishnaraj 1997). Studies of the 1999 and 2002 Turkey earthquakes, the 1999 Taiwan earthquake, the 2010/2011 New Zealand earthquakes, and the 2016 Christchurch, New Zealand earthquake also revealed mortality rates to be elevated among women (Chou et al. 2004; Ellidokuz et al. 2005; Johnston et al. 2014; Lambie et al. 2016). Regarding the 1976 earthquake in Guatemala, a study (Glass et al. 1977) notes that female death rates peaked in the youngest and oldest age groups.

However, researchers have found no strong association in the United States between age and earthquake mortality or gender and earthquake mortality. Furthermore, death rates were similar between males and females in the 1980 Italy and 1999 Taiwan earthquake (Peek-Asa et al. 2003). For the

1980 southern Italy earthquake, de Bruycker and his colleagues (1985) also reported that the death rates by age group did not differ greatly.

The type of building material of residential and nonresidential structures, age, size, height, weight, and density of buildings are generally associated with earthquake-induced mortality (Ramirez and Peek-Asa 2005). Almost all of the housing characteristics, which are an important part of the built environment, have direct relationship with household income. In addition, lack of building codes in earthquake-prone areas is a major contributing factor to deaths.

Some evidence exists that wooden buildings are more easily damaged than reinforced concrete buildings, and hence the former is an important risk factor for earthquake deaths. For example, most deaths in the 1995 Kobe earthquake occurred in the eastern part of the city where most of the houses were old and made of wood with heavy roofs (Masai et al. 2009). Similarly, Ellidokuz et al. (2005) found regarding the 2002 Turkey earthquake that individuals in wooden buildings were at 3.6 times greater risk of death or injury than those in other types of buildings. Earthquake mortality reports in the United States reveal analogous findings (Bourque et al. 2007). This finding was also true for the 2001 Gujarat, India, earthquake (Roy et al. 2002).

Other researchers like Peek-Asa et al. (2003) and Abbott (2008) maintain that lightweight wooden houses, particularly wood tied together well with nails, bolts, and braces, may prove less of a risk than some concrete structures that lose their rigidity. This is likely because such wooden houses handle earthquake shaking quite well as they are flexible and deform without breaking. Conversely, they also stress that wooden houses are susceptible to earthquake-related fires. However, the overall impact of earthquake mortality caused by wooden houses remains inconclusive.

Multistoried residential and commercial buildings are generally thought to cause more fatalities than single-story buildings (Armenian et al. 1997; Masai et al. 2009). However, using data collected after the 1994 Northridge, California, earthquake, Peek-Asa et al. (2003) found no difference in risk of injury to those injured in multiunit residential structures versus those in single-family homes (also see Mathue-Giangreco et al. 2001). This is because the buildings in California strictly follows building codes that make buildings relatively strong and thus less vulnerable to collapse. However, Peek-Asa et al. (2003) did observe a direct association between earthquake fatalities and the age of a building (also see Ramirez and Peek-Asa 2005).

Moreover, proximity to any type of structure during an earthquake is also a risk factor for deaths and injuries (Peek-Asa et al. 2003; Ramirez and Peek-Asa 2005). This helps to explain why deaths are likely following structural collapse of infrastructure such as bridges and multilevel highways (e.g., Eberhart-Phillips et al. 1994; Bourque et al. 2007). For instance, collapse of freeway structures was involved in 81 percent of deaths on public roadways in the Loma Prieta, California earthquake (Eberhart-Phillips et al. 1994). Automobile crash fatalities also are associated with damage to transportation infrastructure, often resulting in disruptions to traffic control devices (Peek-Asa et al. 1998). Like buildings, bridges and roadways are parts of the built environment.

Risk of death from earthquakes does differ between urban and rural areas, however. The risk is higher in urban areas than in rural areas in developed countries for several reasons: Greater density and number of residential and commercial buildings, and more infrastructure, which may collapse and cause more deaths, and of course, more people. This suggests that population and housing density are more likely to result in more deaths in urban areas than in rural areas in developed countries (Peek-Asa et al. 2003). People living in rural areas in developing countries also confront different types of problems: Substandard housing and distance from first responders (Ramirez and Peek-Asa 2005). Note that a relatively high proportion of people in urban areas of developing countries may be forced to live in weak and unsafe structures or substandard facilities that are more likely to collapse during an earthquake (Ramirez and Peek-Asa 2005).

The extent of mortality also varies by level of development of the country struck by an earthquake (e.g., Bourque et al. 2007; Sullivan and Hossain 2009; DesRoches et al. 2011; Elnashai et al. 2011). Because of the typically greater concentration of people, widespread poverty, and poor housing conditions, risk of death is higher in developing countries than in developed countries (Ramirez and Peek-Asa 2005).

In a 2018 study, Paul and his colleagues examined the spatial patterns of death caused by the 2015 Nepal earthquakes to identify the determinants of earthquake mortality at the subnational or district level for the country. Initially, they considered two hazard factors (distance (in km) from the epicenters of the 25 April and the 12 May, earthquakes, to the center of selected district), and six vulnerability factors (number of houses destroyed per 1,000 households, number of houses damaged per 1,000 households, population density, location of residents in terms of ecological zones, literacy rate, and percentage of residents living in urban centers).[4] Application of a spatial Bayesian regression model determined three statistically significant factors (e.g., number of destroyed and number of damaged houses per 1,000 households, and distance from the epicenter of second major earthquake) of 2015 earthquakes deaths in Nepal. All three determinants have predicted direction of association with the dependent variable, that is, the number of deaths per 10,000 population caused by the two major earthquakes.

It was clear that the built environment was the most important determinant of earthquake mortality in Nepal. This was not surprisingly, rather consistent with the popular proverb that "earthquake[s] don't kill people, building[s] do" (Walsh 2010). As indicated, nearly three-quarters of all deaths in earthquakes are caused by building collapse (Cross 2015). Irrespective of population density, literacy rate, and residential location in three ecological zones or rural or urban areas, an overwhelming majority of the houses in Nepal are not built following the seismic codes. This implies that a high proportion of people in Nepal live in unsafe structures or sub-standard facilities that are more likely to collapse during an earthquake. This is an important finding because substantial reduction of earthquake deaths can be achieved in Nepal by only one measure, which is improving the quality of the housing throughout the country.

However, although hazard and vulnerability factors have independent associations with earthquake deaths, when they interact with each other, the risk of deaths is even higher (Ramirez and Peek-Asa 2005). For example, for the same magnitude, seismically stable structures constructed on loose soil may collapse in an earthquake, while less stable structures located on solid rock may remain standing. The former may cause deaths and injuries, but not the latter.

Tsunamis

As indicated in Chapter 3, deaths caused by tsunamis are included with earthquakes primarily because the former events are originated by the latter. Although both share vulnerability factors, hazard factors are different. The extent of death caused by tsunami is determined by elevation and slope of the coast, height and duration of tsunami waves, and shape and configuration of the coast. Tsunami waves originate in the sea and the created waves approach shallow waters along the coasts, bays, or harbors.[5] These waves can travel rapidly inland as far as 20 miles (32 km) from coast, depending on the shape and slope of the shoreline. The triangular shape helps to funnel sea water toward and beyond the coast and causes further amplification of the surge. The speed, height, and strength of the waves determine the damage and deaths. These three physical parameters of tsunami along with shape of the coast in turn determine how far inland water can travel from the coastline. Distance traveled by tsunami waves determines the extent of flooding, and it is very deadly if the waves bring with them destroyed or damaged buildings, trees, and other debris. Onrush of water can also destroy bridges and roads in coastal areas (Aida et al. 2017).

Tsunamis do not consist of one solitary wave, but instead of series of multiple fast-rising and fast-moving waves that bring powerful current with them and often last for hours. People who are injured in the initial wave consequently may be more severely injured or die in subsequent waves. For example, several coastal areas of Sri Lanka experienced at least two waves during the 2004 Boxing Day tsunami. The first wave over three feet (one m) high was followed 10 minutes later by a second wave that was over 30 feet (10 m) high. Ultimately, over 31,000 people died from the tsunami, many because of drowning as a result of subsequent waves (Yamada et al. 2006).

Like other natural disasters, demographic and socioeconomic factors are associated with tsunami deaths. After the 2004 IOT, Oxfam carried out a household survey in four coastal villages in Aceh, Indonesia. This survey reveals that up to four women died for every male in the selected villages. In some villages, all the deceased were women (Oxfam 2005). This happened for several reasons. First, the tsunami waves came to the Aceh coast in early morning where wives of fishermen were waiting for their husbands to return from overnight fishing in the ocean to sort out catches for sale in the markets. This was one of the reasons for the relatively larger number of deaths among women than men, and thus, timing of tsunami is also a risk factor for death. Additionally, many women did not know how to swim or climb trees; because of cultural restrictions on women's behavior, they had not been encouraged to learn to swim or climb trees. As in Aceh, Oxfam also reported that about two-thirds of those who died in the 2004 tsunami were women in the Indian state of Tamil Nadu and in Sri Lanka (also see Nishikiori et al. 2006). Nicole Howe (2019) claimed that 70 percent of all deaths caused by the 2004 IOT were women.

In studying the risk factors associated with mortality in the eastern coastal district of Sri Lanka, Nishikiori et al. (2006) reported that the location of individuals at the time of a tsunami was significantly associated with mortality. One of the reasons for the disproportion of female victims in Sri Lanka was that the tsunami hit the country's east coast at the time when women traditionally bathed in the sea (Howe 2019).

In general, individuals inside a building at the time of tsunami have a significantly higher mortality than those who are outdoors. This is because people who are outside could run to higher ground to escape destructive tsunami waves. In the tsunami-affected countries in 2004, the level of house destruction correlated linearly with mortality; a significantly high mortality was observed among people whose houses were destroyed as compared to those whose property suffered minimum destruction. However, Nishikiori et al. (2006) found no evidence of association between income and mortality. They further reported that children and the elderly had higher mortality rates than adults. A similar finding was also reported by Rofi et al. (2006), who conducted a study on tsunami mortality and displacement in Aceh Province, Indonesia. Nishikiori et al. (2006) also found that tsunami mortality tended to decrease with higher educational level.

Droughts

Droughts (with famines and associated food insecurity) are the second leading cause of death from natural disasters. These are slow onset disasters, and less dramatic than other types of extreme events. Effects of this natural event are not felt at once, which makes it often difficult to determine when one begins and ends. Unlike most other disasters, the study of drought-induced death has been limited because of the unique characteristics of droughts such as gradual onset, persistence, and large geographical extent (Stanke et al. 2013; Berman et al. 2017). Additionally, drought definitions are region- and type (meteorological, agricultural, hydrological, and socio-economic)-specific, which can create problems in assessing risk factors for death. Another challenge of assessing droughts is that droughts mostly kill people through famines. More specifically, droughts affect food production, which, in turn, causes famines.

Droughts can last for even several years or decades. Thus, they slowly take hold in an area and tighten their grip over time. As noted, they are widespread compared to other natural disasters, and they may begin at any time of the year and be only seasonal in certain places. Only three physical characteristics of droughts (duration, areal extent, and magnitude) can be considered as risk factors for deaths.[6] Individually, these characteristics are directly associated with drought deaths. When the magnitude of a drought is high, duration is prolonged, and covers a widespread area; it usually causes high mortality.

At the individual or household level, vulnerability factors of drought deaths include demographic and socioeconomic conditions. Higher mortality rates are found among infants and young children than among older children and adults. Poverty and poor health and nutritional status are also associated with high mortality from drought (Stanke et al. 2013) partly because poverty and poor nutritional status are highly correlated. A study (Sharma 1995) conducted in India demonstrated that drought-induced malnutrition was the root cause of human deaths. Moreover, food intake differs between rural and urban areas. Generally, people of rural areas are more vulnerable to droughts than urban residents because of a higher incidence of hunger, starvation, anemia, and malnutrition. In addition, droughts kill more women than men due to the low status of women in less developed countries. This dictates a traditional food distribution system among the family members in these countries where men eat "first" and "more quantity and quality" than women (Sen 1981).

Cyclones/hurricanes/typhoons

Tropical cyclone-related deaths are a function of magnitude, duration, landfall time, frequency, and spatial extent. This event generally occurs with storm surges that quickly flood low-lying coastal areas and offshore islands at different heights, depending on magnitude. Surge height ranges from 3 feet (1 m) for a category 1 cyclone to over 19 feet (6 m) for a category 5 storm and account for most of the cyclone-related deaths (Paul 2011). Storm surge height is amplified if it coincides with the normal diurnal high tide and/or corresponds with a full moon (Rahman et al. 2015). Surge height is also related to the geography of the coastal environment including width and depth of the shelf. For example, coastal waters that are shallow can amplify storm surge. Consequently, people find it very difficult to swim in a high tidal surge, which rapidly comes inland with high wind, and thus, surge height and speed are directly associated with deaths.

The length of a storm's surge is an important determinant of the number of deaths caused by a cyclone. Moreover, whether the cyclone-affected coastal areas experience only one storm surge instead of a succession of surges also influences the number of deaths. Timing of landfall is also very important for the number of deaths caused by a tropical cyclone. For instance, whether landfall occurs in daytime or nighttime is important because nighttime landfall is more deadly than daytime landfall (Paul et al. 2010).

Cyclone magnitude, which is measured on a five-point scale with a Category 5 cyclone as the most severe and destructive, and Category 1 as the weakest, is generally directly associated with human deaths but not always. For example, Cyclone Sidr, a Category 4 storm, claimed 3,406 lives in Bangladesh in 2007; however, another Bangladeshi cyclone of the same magnitude, Cyclone Gorky, killed 140,000 people in 1991. In 1970, the Category 3 Bhola cyclone killed anywhere from 300,000 to 600,000 people in the same country (Paul 2009).

The relatively low number of deaths caused by Cyclone Sidr was primarily due to implementation of preparedness and mitigation measures. At the time of the Bhola Cyclone, there was no early warning system (EWS) and the concept of evacuation was unknown.[7] After Cyclone Gorky, the Bangladesh government invested in an EWS that provided timely weather forecasting and advance warning systems, resulting in the successful evacuation of people living in coastal Bangladesh. Although construction of public cyclone shelters started after the Bhola Cyclone, there were inadequate numbers of such shelters in the coastal areas. Prior to the landfall of the Cyclone Gorky in 1991, shelters were distributed along the entire coast of Bangladesh, and after that cyclone, more than 2,500 public shelters were constructed (Paul 2009).

Frequency of cyclone is directly related to deaths. Each year, cyclone-prone countries experience multiple storms. Some countries have only one cyclone season, whereas other countries have two seasons. For example, countries of the Northern Indian Ocean (Bangladesh, India, Myanmar, and Pakistan) have two distinct seasons: March through July (Pre-Monsoon) and September through December (Post-Monsoon). It is believed that cyclones that occur in the Post-Monsoon season are more deadly than those in the Pre-Monsoon season.

Cyclone-related deaths are influenced by demographic (age and gender), and socioeconomic (e.g., income, house type, occupation, and educational level) characteristics of affected individuals or households. Individuals' compliance with warning systems, evacuation orders, and taking refuge in designated evacuation centers also determines the number of deaths. A study conducted by Chowdhury et al. (1993) in Bangladesh after Cyclone Gorky reported that 63 percent deaths were children under age 10, who represented only 35 percent of the pre-cyclone population. This study also reported that 42 percent more women died than men. Similar findings were also reported in another study (Bern et al. 1993) dealing with the risk factors of death from the same cyclone. On the other hand, Howe (2019) reported that Cyclone Gorky killed 14 women for every man. She further found from mortuary data that the difference between male and female deaths caused by Hurricane Katrina was negligible. She concluded that where the gender difference is less in countries where economic and social rights were more equally distributed.[8]

Older age is also a risk factor because of underlying health conditions of the elderly as well as their difficulties in receiving and understanding cyclone warnings (Ching et al. 2015). Supporting this theory, Morrow (1999) reported that the elderly were at higher risk of death from Hurricane Andrew, which struck the United States and the Bahamas in 1992, than were younger adults.

Evidence across the world suggests that cyclone deaths are influenced by socioeconomic conditions of coastal residents. For example, Ensor and Ensor (2009) indicated that poor people disproportionately died from Hurricane Mitch in Honduras, which killed between 5,500 and 9,000 people when it made landfall on the Caribbean Sea coast of Honduras in 1998. Using Australian cyclone mortality data for the past four decades (1970–2010) and applying negative binomial regressions, Seo (2015) reported that the lower the income, the larger the number of fatalities from a hurricane. He further found that the hurricane that made landfall in Western Australia and Northern Territory coasts have caused, on average, a larger number of deaths than the cyclones that made landfall in the coasts of Queensland that have higher proportion of affluent people. Morrow (1999) also reported that recent immigrants, woman-headed households, persons in poverty, and persons with special medical needs are at greater risk of cyclone mortality.

Widespread poverty and type of housing are closely related to each other. Usually, houses of the poor are not strong enough to withstand the onslaught of tidal waves; therefore, many coastal residents are killed by collapse of houses. Chowdhury et al. (1993) reported that only 3 percent of residences were strong enough to withstand the storm in coastal Bangladesh when hit by Cyclone Gorky in 1991. The loss of lives from cyclones among fisher folk, who live near the coast, is also greater than for people of other professions (see Frank and Husain 1971).

Among other factors, compliance with evacuation order depends on pet ownership and/or presence of elderly in a household. In Hurricane Katrina, many residents of New Orleans, Louisiana, did not leave their houses because they had either pets or elderly members of the household (Paul 2019). Before landfall of 2007 Cyclone Sidr, people in coastal Bangladesh were reluctant to evacuate to a public cyclone shelter until they had put their livestock in a safer place (Paul and Dutt 2010). One study (Paul 2009) reported no deaths among people who went to a public shelter before landfall of Cyclone Sidr. All deaths occurred among people who did not comply with evacuation orders or who turned back to home because many of the shelters were already full and even over-crowded. In the case of Cyclone Gorky, Bern et al. (1993) claim that 22 percent of persons who did not reach a public cyclone shelter died, whereas all persons who sought refuge in such structures survived. These shelters can withstand storm surges and dangerous winds because they are concrete buildings whose first floor is open for storm surge-induced flooding. Thus, the risk of dying is related to type of shelter and the actions taken to seek shelter.

A similar finding was also reported by Ching et al. (2015) who studied the risk factors for deaths due to Typhoon Haiyan, which made landfall on the central coast of the Philippines in 2013 and killed 6,300 people. Applying a multivariate model, they identified two significant risk factors: Not taking refuge to the designated shelter before the typhoon and exiting the house during the storm. The former appeared to be the greatest risk factor for mortality during the typhoon; however, leaving the house during the storm is also a risk factor because of dangerous winds and water. Furthermore, no knowledge of the approaching storm was an important risk factor determined after Typhoon Bopha, which hit the Philippines in 2013 (Ching et al. 2015).

Tropical cyclone deaths disproportionately befall developing countries. Leading explanations for this difference between developed and developing nations is the size of the vulnerable population and the lack of capacity of pre-event evacuation (Doocy et al. 2013). Thus, GDP per capita is inversely associated with risk of high mortality from cyclones. In developed countries, males are generally associated with increased mortality risk, whereas females experience higher mortality in developing countries. Doocy et al. (2013) further claim that certain WHO regions (Africa, Americas, Europe/ Eastern Mediterranean, South East Asia, and Western Pacific) and even decades (1980, 1990, and 2000) were significantly associated with excess mortality. As a region, South East Asia has recorded more deaths than the other regions and the 2000 decade experienced the highest number of deaths.

Tornadoes

Several host or vulnerability factors (e.g., demographic and socio-economic factors, location (indoor or outdoor), and type of housing, materials, age and height, and type of shelter use) are closely associated with deaths resulting from tornadoes. In addition, there are many physical factors associated with a tornado such as severity, timing of tornado (day and night; also fall and winter months), tornado tracks (the length and width), and duration are determinants of tornado deaths. Provic (2012) maintains that underlying terrain conditions (e.g., land cover, vegetation, surface roughness, and slope) correlate with tornado intensity and hence extent of damage and deaths (also see Wurman et al. 2007). In general, the greater the magnitude of a tornado, the higher the fatality potential (Paul 2011). Simmons and Sutter (2011) state that fatalities per tornado differ by a factor of roughly 15,000 when comparing EF5 tornadoes with EF0 tornadoes. According to them, FE4 and EF5 tornadoes together account for 62 percent of all deaths caused by these events in the United States, yet those tornado ratings make up just 1.2 percent of all occurrences. They also claim that residence in the southeastern states of the United States means a higher risk of death from tornadoes than in other states.

Tornadoes occurring at night are more than twice as likely to be deadly as those during the day (Ashley 2007) for two reasons: At night people tend to stay home, and second, they are often asleep. Both situations preclude hearing the tornado siren and being able to take protective shelter. Another reason is related to fewer storm spotters at night, reducing the ability of meteorologists to confirm nighttime tornadoes in progress. Strong tornadoes can occur in any month, but April, May, and June are the most dangerous months for tornadoes in the United States (Simmons and Sutter 2011).

Several researchers (e.g., Karstens et al. 2013; Franklin et al. 2015) suspect a relationship between elevation and number of tornado deaths. Stimers and Paul (2017) examined deaths caused by the 2011 Joplin tornado partitioned by elevation of the damage path. This path stretched a total of 22.1 miles (35.6 km). Six miles (9 km) of the total track crossed through the city of Joplin, whose elevation is approximately 164 feet (50 m) from beginning to end. They reported that the elevation and tornado fatalities are inversely related, but the relationship is not statistically significant. This lack of a significant relationship was explained in terms of tree coverage. Such coverage around buildings was negatively associated with tornado fatalities in Joplin. Thus, for most of the deaths that occurred in nonresidential buildings in the city, those buildings were not surrounded by trees. Tree surroundings create barrier for wind movement and thus reduce the intensity of the tornado (Stimers and Paul 2017).

The number of deaths is directly associated with the length and width of a tornado tract and its magnitude (Simmons and Sutter 2011, 2012).[9] The strength of tornadoes varies not only linearly but also within or across the path. This is most evident when the EF scale is applied along the tornado tract where it most accurately represents the event's magnitude at the central zone of the tract. This zone experiences the most damage and is the location of the most deaths. The strength of the tornado decreases toward the two outer zones of the path adjacent to the central zone. Thus, the outer zones are less deadly than the central zone. Most studies of tornado deaths ignore this fact and consider a tornado path as a line without a contributing width. However, in studying the 2011 tornado mortality, Paul and Stimers (2014) found that the number of deaths systematically decreases with increasing distance from the central zone. They further reported that 122 of the 154 deaths (79 percent) occurred in the central zones. Applying a chi-square test, they reported that the number of deaths statistically decreased outwardly from the central zone (Paul and Stimers 2014).

Moreover, if the tornado track travels over densely populated areas, death toll is likely to be higher than if it travels over less populated areas (Balluz et al. 2000; Wurman et al. 2007). While studying the tornado events in the United States for the 2000–2009 time period, Stimers and Paul (2016) found that the percent of tornadoes that struck communities or places differed remarkably from state to state, ranging from a minimum of 1.73 percent (Colorado) to a maximum of 25 percent (New Hampshire) with an average of 7.75 percent for the country as a whole. However, in contrast to population density, housing density tends to reduce mortality risk by providing friction-based wind resistance that would mitigate violent winds.

Also, age of residents is a risk factor for such deaths (Simmons and Sutter 2011). In the 2011 Joplin, Missouri, tornado, approximately eight fatalities occurred per 1,000 people aged 60 years and over compared with two fatalities per 1,000 people below 60 years (Kuligowski et al. 2013). Similarly, other studies (e.g., Edison et al. 1990; Kunii et al. 1996; Sugimoto et al. 2011) in the United States and Bangladesh demonstrated that the elderly were at increased risk of death or injury from tornadoes. Old people are more vulnerable to death because of weaker bodies, limited movement ability, partial or full hearing loss, and/or living alone (Deng et al. 2019). Moreover, many elderly may not have access to tornado warnings, and if they do, they also have a tendency to dismiss them (Schmidlin and King 1995). Females are also more vulnerable than males (Chiu et al. 2013; Sugimoto et al. 2011),[10] and weak bodies are more susceptible to the collapse of buildings and blunt force trauma from heavy objects that are airborne during a tornado event. However, risk of death from tornado can be reduced significantly by staying away from windows, doors, and outside walls (Edison et al. 1990).

Regarding type of residence, houses with glass windows, wooden houses, trailers, or mobile car-houses increase the likelihood of death (Bohonos and Hogan 1999; Schmidlin and King 1995). Moreover, wooden homes seem to offer less protection than brick homes. Curtis and Fagan (2013) reported that those in multiple-level apartment residences have a higher probability of death from tornadoes than those in one-level and single-family residences. Particularly, people are often struck by broken window glass or other falling objects. Note that flying objects and collapse of buildings are the two major causes of tornado deaths. Staying outside in a tornado also increases the risk of death. Studies by Carter et al. (1989) and Daley et al. (2005) reported that being outdoors during tornadoes was associated with significantly elevated risk of death when compared to being indoors. The magnitude of the OR for being outdoors was 141.2 and infinity, respectively. Conversely, sheltering in a sturdy house decreases such risk as can advance tornado warnings (Schmidlin and King 1995; Bohonos and Hogan 1999; Daley et al. 2005). Predictably, the relative risk (RR) of death for mobile or manufactured home occupants was 11.5 times the risk for conventional home occupants (Edison et al. 1990).

While studying risk factors from tornadoes in China, Deng et al. (2019) reported that a higher household income was associated with lower odds of injury and death. Such income level may be associated with owning sturdy houses. They further indicated a close inverse relationship between household income and the degree of housing collapse. In this context, it should be noted that tornado deaths are also directly associated with age of the house. Old houses built decades ago are liable to collapse because they are not built with modern tornado-resistant technology. Previously, Paul and Stimers (2012) had found after the 2011 Joplin, Missouri, tornado that older houses were at greater risk of collapsing. In addition, Bohonos and Hogan (1999) reported that high-rise buildings were more deadly in tornadoes than low-rise buildings in the United States. A similar finding is also reported by Deng et al. (2019) in their China study. Accordingly, sheltering in the basement is negatively associated with risk of tornado deaths. This helps explain one of the reasons for the record number of deaths in Joplin, Missouri, tornado in 2011. The ratio of the city houses that had a basement to those that did not was very low (Paul and Stimers 2012).

Blizzards and ice storms

EM-DAT considers blizzards and ice storms under the broad category of storms, which combined contributed to nearly 20 percent of the disaster deaths during the 1991–2015 period. These two storm events are considered together because they share determinants of deaths. Like other hazard and disasters, winter storms' unique characteristics (e.g., strong winds, high amounts of snowfall, frigid temperature, heavy blowing snow, frequency, and duration) are positively correlated with deaths. Two other physical dimensions (seasonality and diurnal factor) are irrelevant as risk factors for winter storm-related deaths.

At the individual or household level, vulnerability factors include age, gender, and location at the time of blizzards and winter storms. A study (Dixon et al. 2005) conducted in the United States showed that males between 40 and 49 years old were most vulnerable to blizzards and other winter storms. The second most vulnerable group was females 20–29 years old. The number of deaths is also high for school students who attempted to walk home in winter storms. Not surprisingly, death rates are higher for people who routinely spend hours a day outside in cold weather; three-fourths of winter weather-related deaths occur outdoors, including in vehicles.

Lightning

Worldwide lightning causes approximately 24,000 fatalities per year with the United States averaging 43 reported lighting deaths per year during the 1989–2018 period. More recently, in the last 19 years (2009–2018), the United States has averaged 27 lightning deaths annually (Jensen and Vincent 2019). In contrast, only two people were killed by lightning in the United Kingdom in 2014 (Elsom and Webb 2015). Of all the physical dimensions of lightning, diurnal and seasonal factors are mostly associated with this event. Although it can occur at any time of the year, at any time of the day, or on any day of the week, it is more frequent in the summer season, during the afternoon, and on a weekend day. More than 70 percent of lightning deaths occur in the United States during the months of June, July, and August when people spend more time outside their homes. In this country, the greatest number of fatalities occurs on weekend days, particularly on Saturdays (Holle 2016).

Lightning does not always strike only tall objects; also, it strikes the ground in open fields or spaces. Therefore, location in terms of outdoors is a major risk factor for lightning deaths. Moreover, outside activities differ between developed and developing countries, which can be associated with lightning deaths. Outside activities in developed countries include fishing, boating, swimming, walking, running, biking, playing golf, baseball, and football, or relaxing at a beach or lake. In the United States, almost two-thirds of lightning deaths are associated with leisure activities. In contrast, in developing countries, farmers and agricultural laborers have a higher risk of deaths from lightning than do those in other occupations. In Mongolia, almost 100 percent of deaths from lightning have been reported for people either in an open field or inside a remote isolated buildings, locally called *yurt* or *ger* (Daljinsuren and Gomez 2015). A large section of the population in Mexico, the country with the worst lightning death rate in the Americas, most at risk of dying while working outdoors is subsistence farmers, including the indigenous population. Thus, most of the deaths are in rural areas. Moreover, construction workers of both developed and developing countries are equally vulnerable to lightning deaths.

The risk of lightning deaths increases if a person is under an isolated tree or holding or conducted to something metallic because metal attracts lightning. In all countries most vulnerable to lightning, more males than females die from this event. One report (Jensenius 2019) claims that in the United States, male deaths accounted for about 80 percent of all lightning deaths during the 2006–2018 period. Another study (Duclos et al. 1990) claims that 87 percent of all lightning deaths in Florida for 1978–1987 were males. The greatest number of fatalities occurs in the United States for people between ages 10 and 60, with a relative minimum number of deaths in the 30–39 cohort. In particular, boys aged 10–19 are the most common victims in developing countries as they generally spend more time outdoors. Lightning disproportionately kills the poor or people who do not have a safe house. Predictably, living in a thatched roof house increases the risk of deaths from lightning.

Floods

A host of physical characteristics of floods can be considered as important determinants of mortality induced by these events. These characteristics include magnitude (depth), duration (short, medium, and long), frequency, scale (spatial extent), water current (still, moderate, and rapid), and type of floods (e.g., coastal, river, and flash). Mortality is thought to increase with any increase in all of these physical characteristics (e.g., Ashley and Ashley 2008; Doocy et al. 2013; Hu et al. 2018; Paul 2011; Paul and Mahmood 2016). A common measure of flood magnitude is the maximum discharge of water at a given point, or how much water in the riverbed is flowing past a certain point in a given period of time.[11] In the United States, flood stage is widely used as a measure of flood magnitude. It is an arbitrary fixed gauge height at which water level of a river overtops its bank (Paul 2011).

Flood depth is often used to represent flood magnitude. In general, flood fatalities and flood depth are directly related to each other. For example, Boyd (2010) applied a univariate regression model with water depth as the independent variable and the flood fatality rate for each flooded census block of New Orleans as a dependent variable. The flood was caused by Hurricane Katrina. However, he found that the flood mortality rate steadily increases as flood depth increases.

Meanwhile, another physical dimension, namely flood frequency, is also directly related to flood mortality rate. Flood frequency is the probability of occurrence of a given magnitude flood.[12] However, physical dimensions of floods interact with each other, leading to deaths. For example, a flood that affects a large area may not cause any deaths because of its low magnitude and/or short duration. In contrast, a short duration but high magnitude flood can increase death risk significantly (Hofer and Messerli 2006).

Flash floods affect a relatively small number of people but can cause relatively high mortality. On the other hand, river floods affect a greater number of people, but result in relatively low mortality. In the United States and European countries, more people die from flash floods than any other type. This is because people, particularly in the United States, drive cars in flash floods without any knowledge of their destructive powers (Jonkman and Kelman 2005). Even six inches (154.4 mm) of rushing water can be difficult to stand in and cause vehicles to become unresponsive or stall, and a couple of feet of water will float cars, or any other large objects.

At the individual or household level, vulnerability factors leading to deaths in floods include age, gender, occupation, disability, income, educational level, clothing worn, swimming ability, past experience, and risk taking behaviors like walking, boating, and playing in floodwater (Ahern et al. 2005; Nishikiori et al. 2006; Pradhan et al. 2007; FitzGerald et al. 2010; Tantiworawit et al. 2012). Particularly, working age (31–60 years) people are at significantly increased risk of death given that children and the elderly stay home, while adults are more likely to expose themselves to floodwater during their daily activities.

This relationship between age and flood deaths is observed in many countries (Tantiworrawit et al. 2012); however, children under 10 years are at a higher risk of death in countries such as Australia, Bangladesh, and Nepal (Bern et al. 1993; Pedan and Franklin 2019; Pradhan et al. 2007). Such children died mainly from drowning as they played or walked in floodwaters with a strong current. A study (Ahmed et al. 1999) in Bangladesh identified a strong association between the annual flooding and child drowning. Another study (Pradhan et al. 2007) in Nepal identified children as recording flood-related fatality rates six times higher than mortality rates in the same village prior to flood. Data from the United States suggest young adults (aged 10–19) have a higher vulnerability to flooding (Ashley and Ashley 2008).

Gender-specific risk of flood mortality has been inconclusive (Doocy et al. 2013). In contrast with the prevailing notion that women are more vulnerable in disasters, Tantiworrawit et al. (2012) found in Thailand that males are significantly more at risk of death because they are involved in water-related activities such as harvesting crops in floodwater, fishing, boating, and walking through floodwater. A study (FitzGerald et al. 2010) in Australia reported that most males died from drowning, while females died from structural collapse. Thus, many flood fatalities among males are associated with risk-taking behaviors.

Like overall disaster mortality rates, flood fatalities are higher among the poor and illiterate than among the wealthy and educated individuals or households. Ahern et al. (2005) reported that floods with the highest mortality have been the cause of death for people with limited resources. Similarly, Pradhan et al. (2007) found in Nepal that those living in a thatched house had a 5.1-fold higher risk of death than residents living in a cement/brick house. Clearly, materials used to make a house represent the socioeconomic conditions of the household owners. However, swimming ability and past flood experience are thought to be negatively associated with mortality. Conversely, clothing, particularly by females in South Asia, is a risk factor because they wear *sari*, which is very long and difficult to swim in (Ikeda 1995).

At the country or other areal unit levels, flood-related mortality are associated with population density, level of economic development, rate of urbanization, type of governance and government, and terrain and geographic characteristics (Albala-Bertrand 2003; Kahn 2003, 2005; Toya and Skidmore 2006; Doocy et al. 2013). Accordingly, flood deaths are generally concentrated in developing and heavily populated countries because of differences in terms of level of development, which is in keeping with the established relationship between poverty and increased disaster deaths. Thus, many studies (e.g., Padli et al. 2013; Hu et al. 2018) found that lower GDP per capita was linked to higher flood mortality. Padli et al. (2013) also claim that countries with higher GDP are more prepared to face future devastation due to floods. Thus, enhancing economic development can reduce the impact of flood on human deaths.

Because developing countries have proportionately more vulnerable populations than do richer nations, vulnerability factors at different levels are interrelated. "Additionally, events of equal magnitude generally pose very different threats for countries at different levels of (economic) development"

(Paul and Mahmood 2016, 1707). Note that hazard and vulnerability factors work either individually or in combination with one another, but often, these two types of factors overlap to some extent. Also, it should be stressed that national- and other areal units-level vulnerability factors are appropriate only for cross-national studies. For temporal studies at a particular areal unit level, many vulnerability factors do not vary within a short period of time, and hence these variables will show little or no yearly variations.

Extreme temperatures

Both extreme heat and cold have been associated with deaths, but the former is more deadly than the latter. Determinants of deaths caused by extreme cold have already been discussed, and so extreme heat is the concerned of this section. Globally, heat waves, hot spells, or extreme heat in general kill many more human than do cyclones, floods, tornadoes, and lightening. In fact, extreme heat events are classified among the 10 deadliest disasters (Guha-Sapir et al. 2012). Moreover, a particular heat event can kill people in very large numbers. For example, the 2003 heat wave in Europe killed over 40,000 people, while in the United States, almost 10,000 deaths were attributed to excessive heat exposure between 1979 and 2001 (Coppola 2006). In recent years, the number of heat-related deaths per year has averaged around 800. This is because the number of hot days and nights has increased in the United States as well as around the world, and this is associated with climate change (Stocker 2014).

Clearly, determinants of mortality during heat waves are not only physical characteristics of the events such as duration, severity, and frequency. Demographic characteristics, pre-existing chronic health conditions and mental disorders, socioeconomic status, and living conditions undoubtedly play a major role in mortality. For example, people over the ages of 65 years exhibit disproportionately higher mortality during such waves than do younger individuals. Elderly people, even in the absence of overt cardiovascular disease, are the most vulnerable to heat exposure (Keatinge 2003; Kenney et al. 2014; Compbell et al. 2018). In the United States, people over 65 years have been several times more likely to die from heat-related cardiovascular disease than the general population, while African-Americans have had higher than average rates (Madrigano et al. 2015).

Apart from the elderly, children and economically disadvantaged groups, including homeless people, are also vulnerable to extreme temperatures. Furthermore, odds of heat wave-related deaths are increased if one is immobile or disabled, lives isolation, is on medication, sleeps on the top floor of a building, and/or is in a structure that lacks thermal insulation (Naughton et al. 2002; Vandentorren et al. 2006). Patrick Abbott (2008) reported that during the 1995 heat waves in Chicago, more than 50 percent of the deaths occurred among those who lived on the top floor of their apartments, where heat buildup was greatest. Thus, poverty and living in large urban areas both in developed or developing countries strongly determined who will die from heat waves. Other identified subgroups with increased vulnerability include those working outdoors or in noncooled environment (Hanna et al. 2011; Yin and Wang 2017).

Irrespective of developed and developing countries, living in large cities greatly exacerbates the heat problem because built up areas are subject to heat island effect. Because of the high concentration of buildings, and paved surface, cities absorb more heat in the day and radiate more heat at night than rural areas. As a result, large urban centers generally experience a lot less cooling at night than do rural areas (Angel n.d.). Kenney et al. (2014) also reported that people living in an underprivileged region or in an unsafe community experience higher risk of deaths than those who live in privileged areas. This was observed in Paris during the French heat wave of 2003, which killed 15,000 people.[13] People having access to transportation and having nearby social contacts are negatively associated with heat wave (Keatinge 2003; Vandentorren et al. 2006). Preparedness can also affect mortality. For example, Chicago experienced two meteorologically similar heat waves in 1995 and 1999, but the latter caused excess deaths of 114 people compared to 700 in 1995. This difference is attributed to increased public awareness and improved response by the city authorities.

Heat waves and drought conditions contribute to wild or forest fires because the extreme heat dries out vegetation, creating tinder for fires. However, globally, wildfires kill relatively fewer people than do other natural disasters. This is primarily for two reasons: Each fire site affects a relatively small area; second, the site is often located in a remote and isolated place. Additionally, unlike other natural disasters, a limited number of countries (e.g., Australia, European countries, and the United States) are vulnerable to wildfires. Nevertheless, those who die from wildfires include the young, the elderly, those with mobility issues, those with pulmonary and respiratory problems or other illnesses, and smokers. The other groups vulnerable to wildfire deaths are workers of forest and fire departments, including firefighting brigades. The 2009 Black Saturday Bushfires in Australia claimed 173 lives. Around 50 percent of these deaths were children younger than 12, people older than 70, or those who had physical disabilities (Cruz et al. 2012). Similar findings are also reported for the United States.

Landslides

The worldwide annual death toll due to landslides is in the thousands. Globally, slightly over 4,000 people were killed by landslides in 2017, while according to the United States Geological Survey (USGS), landslides kill between 25 and 50 people each year in the United States. However, there is a dearth of studies on risk factors for death associated with landslides (Kennedy et al. 2015). One study (Sanchez et al. 2009) reported that females had a higher mortality rate from landslides in Chuuk, Federated States of

Micronesia than males, and children aged 5–14 years. The same findings were also reported by Agrawal et al. (2013), who studied risk factors for mortality in landslides and flood affected population in Uganda.

Sanchez at al. (2009) further observed that people being aware that landslides had recently occurred in the country and knowledge of natural warning signs were significantly associated with lower risks of deaths. Being inside a building during landslides also increased mortality, but this was not statistically significant. The study found no association between the size of a landslides or the slope angle with human mortality. Finally, another study (Guzzetti et al. 2005) conducted in Italy reported that rapid landslides events (e.g., debris flows and mudflows) were more dangerous for killing people than slower moving events such as rotational landslides.

Volcanoes

Distance is the most important factor for volcanic deaths: The closer one is to a volcano, the greater the odds of dying if it erupts. Thus, local residents have the highest risk of death, as do tourists, volcanologists, and members of the media. These groups dominate the proximal fatality record up to 3.3 miles (5 km) from a volcano, typically. Brown et al. (2017) claim that more deaths occur within the first 3.3 miles (5 km) of the erupting volcano than any other 3.3 miles (5 km) zone beyond. For large eruptions, Auker et al. (2013) further claim that over 90 percent of volcano-induced deaths occur between 3.3 miles (5 km) and 20 miles (30 km).

Most people killed by volcanoes are the victims of pyroclastic flows (mixtures of hot gas and ash), which can reach a distance of several miles and tens of miles in large eruptions (Walker 1983). These flows move too rapidly for people to escape, and death is almost certain. Baxter (1990) claims that deaths are commonly caused by thermal injury, asphyxiation, and blast trauma. He further claims that death and injury ratios reach as high as 230:1. Apart from pyroclastic flows, people also die from tsunamis, lahars (volcanic mudflows), tephra (fragmental materials of any size and origin produced a volcanic eruption), ballistics (large ejected clasts), avalanches, lava flows, gas (e.g., carbon dioxide, hydrogen sulfide, sulfur dioxide, and carbon monoxide), and volcanic lightening. Tsunamis can result from the rapid entrance of debris avalanches and other volcanic products into a large water body (Brown et al. 2017). Auker et al. (2013) added pyroclastic density as a correlate of volcanic deaths.

Eruption size or strength is directly associated with the number of deaths. However, larger eruptions are relatively infrequent. Eruption size is often called the Volcanic Explosivity Index (VEI). The index is based upon the volume of tephra and other types of ejected materials produced during an eruption. The VEI scale is open-ended with the largest volcanic eruptions in history being given magnitude 8. A value of 0 is assigned for nonexplosive eruptions. The scale is logarithmic, with the exception of VEI0, VEI1, and

VEI2. The available data show a tendency for fatality numbers to increase with VEI (Auker et al. 2013).

The more people who live around a volcano, the greater the probability of death from an eruption. Thus, population density around volcanoes is correlated with deaths. Additionally, if volcanic ash builds up on roof tops, it often leads to collapse of houses. For example, most of the 350 deaths caused by the eruption of Mount Pinatubo, in the Philippines, in 1991 were due to collapsing roofs.

Tephra can also destroy crops, leading to widespread crop failure. This can result in famine, the largest indirect hazard of volcanic eruptions (Oppenheimer 2003). In 1815, after the eruption of Mount Tambora, Indonesia, 80,000 people died due to famine (Riley n.d). Associated famine was also reported in Europe. Indirect deaths can also occur because of vehicle accidents associated with slippery and hazy road conditions caused by ash. These accidents result from fleeing or evacuating the volcanic site. Thus, indirect deaths can occur not only over great distance but also over a considerable time period after an eruption (Brown et al. 2017).

Conclusion

By a careful and systematic review of existing literature, this chapter has discussed the important determinants of deaths caused by various natural disasters. This information is important for implementation of appropriate preparedness and mitigation measures to reduce mortality level from extreme events. For example, the compliance with evacuation orders is still low in cyclone-prone countries like Bangladesh and the Philippines. These countries also suffer from an inadequate number of public cyclone shelters and therefore many coastal residents have returned home from such facilities because of overcrowded conditions. Clearly, building more public shelters would significantly reduce deaths from cyclones. Developing or spreading awareness through public education is also useful to decrease mortality from specific types of disasters. For example, people both in developed and developing countries generally seek shelter under isolated trees or tall objects because they believe erroneously that such trees and objects provide protection from lightning. Moreover, often, they seek shelter under a tree to avoid rain (Duclos et al. 1990).

The systematic review of determinants of different disasters clearly suggests that the elderly is the most vulnerable group for all disasters. This group should be protected because the global population of the aged is rapidly growing in both developed and developing countries. Without protection, the number of people at risk of dying from future natural disasters tends to increase exponentially. Other groups at high risk for all natural disasters are children, women, the disabled, and the poor. They should also attract special attention of hazards and disasters managers, policy makers, NGOs, donor agencies, and other multinational agencies and humanitarian organizations.

Notes

1. Two associated terms are relative risk (RR) and absolute risk (AR). The RR is a ratio of the probability of an event occurring in the exposed group versus the probability of the event occurring in the nonexposed group. The AR is the probability or chance of an event, that is, the number of events that occurred in a group, divided by the number of people in that group.
2. The influence of these physical dimensions on deaths is discussed in the subsequent sections under each type of disaster.
3. The process of liquefaction occurs when ground shaking causes water to rise, filling pore spaces between granular sediments, increasing pore water pressure and resulting in the sediment acting as a fluid rather than a solid.
4. Ecologically, Nepal is divided into three zones: Northern mountains, central hills, and southern terai or lowlands. Because of higher elevation, districts located in the northern and central zones are more susceptible to both earthquakes and earthquake-induced landslides than the districts of the southern lowlands (Khazai et al. 2015).
5. Note that in deep waters, waves created by tsunami are relatively unnoticed, often felt by those in the sea as a gentle wave (Nishikiori et al. 2006).
6. Drought magnitude is widely expressed as the Palmer Drought Severity Index (PDSI). It is a measure of both the intensity and magnitude of any drought. The PDSI uses a value of "0" to show normal conditions while negative values are representative of drought conditions. The standardized index ranges from −10 (dry) to +10 (wet) (Paul 2011).
7. Cyclone Nargis made landfall in Myanmar in 2008 and killed 138,366 people. This event killed so many people because the Myanmar government was not prepared for this disaster. It neither issued a cyclone warning nor evacuated a single person from the potential affected areas (Asia-Pacific Center 2008).
8. In both developed and developing countries, women and girls are at greater risks than their male counterparts from postdisaster violence and exploitation. Howe (2019) reported that after Haiti earthquake in 2010 many women and girls were targets of armed gang. She further reported that the rate of gender-based violence toward women in Mississippi State increased from previous year more than threefold after landfall of Cyclone Katrina in 2005.
9. The length of a tornado track can be up to 100 miles (150 km). Width of tornado tracks also varies, ranging from a few feet to a mile (1.5 km) or more (Paul and Stimers 2014).
10. Paul and Stimers (2012) did not find significant variations by gender in the 2011 Joplin, Missouri, tornado deaths. Also, in a Bangladeshi study, Sugimoto et al. (2011) reported that tornado deaths did not vary significantly by gender after adjustment for differences in age and location during the tornado.
11. In some countries, such as Bangladesh, flood magnitude is often expressed as the maximum height reached by flood waters above sea level, or simply above ground (Paul 2011).
12. Flood frequency is often expressed as a return period. For example, the return period of a flood might be 100 years. It is expressed as a return period of 1 percent in any given year. Note that some countries experience more than one flood in a given year.
13. These huge number of deaths led to a shortage of space to store dead bodies in mortuaries. Temporary mortuaries were set up in refrigerator lorries.

References

Abbott, P.L. 2008. *Natural Disasters*. New York, NY: McGraw Hill.

Agrawal, S., Y. Gorokhovich, and S. Doocy. 2013. Risk Factors for Mortality in Landslides- and-Flood-Affected Populations in Uganda. *American Journal of Disasters* 3(8): 113–122.

Ahern M., R.S. Kovats, P. Wilkinson, R. Few, and F. Matthies. 2005. Global Health Impacts of Floods: Epidemiologic Evidence. *Epidemiologic Reviews* 27: 36–46.

Ahmed, M.K., M. Rahman, and J. van Ginneken, 1999. Epidemiology of Child Deaths due to Drowning in Matlab, Bangladesh. *International Journal of Epidemiology* 28: 306–311.

Aida, J., H. Hikichi, Y. Matsuyama, Y. Sato, T. Tsuboya, T. Tabuchi, S. Kayama, S.V. Subramanian, K. Kondo, K. Osaka, and I. Kawachi. 2017. Risk of Mortality During and After the 2011 Great East Japan Earthquake and Tsunami Among Older Coastal Residents. *Scientific Reports* 7: 16591 (nature.com/articles/s41598-017-16636-3#Sec4).

Albala-Bertrand, J.M. 2003. *Political Economy of Large Natural Disasters: With Special Reference to Developing Countries*. Oxford: Clarendon Press.

Alexander, D. 1993. *Natural Disasters*. New York, NY: Chapman and Hall, Inc.

Angel, J. n.d. The 1995 Heat Wave in Chicago Illinois (www.isws.illinois.edu/statecli/General/1995Chicogo.htm – last accessed April 1, 2020).

Armenian, H.K., E.K. Noji, and A.P. Oganesian. 1992. A Case-Control Study of Injuries Arising from the Earthquake in Armenia, 1988. *Bulletin of World Health Organization* 70(2): 251–257.

Armenian, H.K., A. Melkonian, E.K. Noji, and A.P. Hovanesian. 1997. Deaths and Injuries due to the Earthquake in Armenia: A Cohort Approach. *International Journal of Epidemiology* 26(4): 806–813.

Ashley, W.S. 2007. Spatial and Temporal Analysis of Tornado Fatalities in the United States: 1880–2005. *Weather Forecasting* 22: 1214–1228.

Ashley, S.T., and W.S. Ashley. 2008. Flood Fatalities in the United States. *Journal of Applied Meteorology and Climatology* 47: 805–818.

Asia-Pacific Center. 2008. *Cyclone Nargis and the Responsibility to Protect*. Brisbane, Australia: The University of Queensland.

Auker, M.R., R. Stephen, J. Sparks, L. Siebert, H.S. Crosweller, and J. Ewert. 2013. A Statistical Analysis of the Global Historical Volcanic Fatalities Record. *Journal of Applied Volcanology* 2(2). https://doi.org/10.1186/2191-5040-2-2.

Balluz, L., L. Schieve, T. Holmes, S. Keiezak, and J. Malilay. 2000. Predictors of People's Response to a Tornado Warning, Arkansas, 1 March 1977. *Disasters* 24: 71–77.

Baxter, P.J. 1990. Medical Effects of Volcanic Eruptions: 1. Main Causes of Death and Injury. *Bulletin of Volcanology* 52: 532–544.

Berman, J., K. Ebisu, R.D. Peng, F. Dominici, and M.L. Bell. 2017. Drought and Risk of Hospital Admissions and Mortality in Older Adults in Western USA from 2000–2013: A Retrospective Study. *Lancet Planet Health* 1: e17–e25.

Bern, C., J. Sniezek, G.M. Mathbor, M.S. Siddiqi, C. Ronsmans, A.M. Bennish, E. Noji, and R.I. Glass. 1993. Risk Factors for Mortality in the Bangladesh Cyclone of 1991. *Bulletin of World Health Organization* 71(1): 73–78.

Blackburn, J.K., T.L. Hadfield, A.J. Curtis, and M.E. High-Jones. 2014. Spatial and Temporal Patterns of Anthrax in White-tailed Deer, Odocoileus Virginianus, and Hematophagous Flies in West Texas during the Summertime Anthrax Risk Period. *Annals of the Association of American Geographers* 104(5): 939–958.

Bohonos, J.J., and D.E. Hogan. 1999. The Medical Impact of Tornadoes in North America. *Journal of Emergency Medicine* 17(1): 67–73.

Bourque, L.B., J.M. Siegel, M. Kano, and M.M. Wood. 2007. Morbidity and Mortality Associated with Disasters. In *Handbook of Disaster Research*, edited by Rodriguez, H., E. Quarantelli, and R.R. Dynes, pp. 97–112. New York, NY: Springer.

Boyd, E.C. 2010. Estimating and Mapping the Direct Flood Fatality Rate for Flooding in Greater New Orleans Duro to Hurricane Katrina. *Risk, Hazards & Crisis in Public Policy* 1(3): 91–114.

Brown, S.K., S.F. Jenkins, R. Stephen, H. Odbert, and M.R. Auker. 2017. Volcanic Fatalities Database: Analysis of Volcanic Threat with Distance and Victim Classification. *Journal of Applied Volcanology* 6, 15. http://doi.org/10.1186/s13617-0067-4.

Brumbaugh, D.S. 1999. *Earthquakes: Science and Society*. Upper Saddle River, NJ: Prentice Hall.

Card, N.A. 2012. *Applied Meta-Analysis for Social Science Research*. New York, NY: The Guilford Press.

Carrel, M., and M. Emch. 2013. Genetics: A New Landscape for Medical Geography. *Annals of the Association of American Geographers* 103(6): 1452–1467.

Carter, A.O., M.E. Millson, and D.E. Allen. 1989. Epidemiologic Study of Deaths and Injuries due to Tornadoes. *American Journal of Epidemiology* 130(6): 1209–1218.

Ching, P.K., V.C. de los Reyes, M.N. Sucaldito, and E. Tayag. 2015. An Assessment of Disaster-Related Mortality Post-Haiyan in Tacloban City. *Western Pacific Surveillance and Response Journal* 6(31): 34–38.

Chiu, C.H., A.H. Schnall, C.E. Mertzlufft, R.S. Noe, A.F. Wolkin, J. Spears, M. Casey-Lockyer, and S.J. Vagi. 2013. Mortality from a Tornado Outbreak, Alabama, April 27, 2011. *American Journal of Public Health* 103(8): e52–e58.

Chou, Y.J., N. Huang, C.H., Lee, S.L. Tsai, L.S. Chen, H.J. Chang. 2004. Who is at Risk of Death in an Earthquake? *American Journal of Epidemiology* 160: 688–695.

Chowdhury, A.M.R., A.U. Bhuyia, A.Y. Choudhury, and R. Sen. 1993. The Bangladesh Cyclone of 1991: Why So Many People Died? *Disasters* 17(4): 292–304.

Compbell, S., T.A. Remenyi, C.J. White, and F.H. Johnston. 2018. Heatwave and Health Impact Research: A Global Review. *Health & Place* 53: 210–218.

Coppola, D.P. 2006. *Introduction to International Disaster Management*. Boston, MA: Elsevier.

Cross, R. 2015. Nepal Earthquake: A Disaster that Shows Quakes does not Kill People, Buildings Do. *The Guardians*, 30 April.

Cruz, M.G., A.L. Sullivan, J.S. Gould, N.C. Sims, A.J. Bannister, J.J. Hollis, and R.J. Hurley. 2012. Anatomy of a Catastrophic Wildfire: The Black Saturday Kilmore East Fire in Victoria. *Forest Ecology and Management* 284: 269–285.

Curtis, A.J., and W.F. Fagan. 2013. Capturing Damage Assessment with a Spatial Video: An Example of a Building and Street-Scale Analysis of Tornado-Related Mortality in Joplin, Missouri, 2011. *Annals of the American Association of Geographers* 103(6): 1522–1538.

Daley, W.R., S. Brown, P. Archer, E. Kruger, F. Jordan, D. Batt, and S. Mallonee. 2005. Risk of Tornado-Related Death and Injury in Oklahoma, May 3, 1999. *American Journal of Epidemiology* 161(12): 1144–1150.

Daljinsuren, M., and C. Gomez. 2015. Lightning Incidents in Mongolia. *Geomatics, Natural Hazards and Risk* 6(8): 686–701.

de Bruycker, M.D. Greco, M.F. Lechat, I. Annino, N., de Ruggiero, and M. Triassi. 1985. The 1980 Earthquake in Southern Italy – Morbidity and Mortality. *International Journal of Epidemiology* 197: 113–117.

Deng, Q., Y. Lv, F. Zhao, W. Yu, J. Dong, and L. Zhang. 2019. Factors Associated with Injuries among Tornado Victims in Yancheng and Chifeng, China. *BMC Public Health* 19, 1556. https://doi.org/10.1186/s12889-019-7887-6.

DesRoches, R., M. Comerio, M. Eberhard, W. Mooney, and G.I. Rix. 2011. Overview of the 2010 Haiti Earthquake. *Earthquake Spectra* 27 (S1): S1–S21.

Dixon, P.G., D.M. Brommer, B.C. Hedquist, and R.S. Cervency. 2005. Heat Mortality Versus Cold Mortality: A Study of Conflicting Databases in the United States. *Bulletin of the American Meteorological Society*, July. http://doi.org/10.1175/BAMS-86-7-937.

Donaldson, A., and D.M. Wood. 2008. Avian Influenza and Events in Political Biography. *Area* 40(1): 128–130.

Doocy, S., A. Daniels, S. Murray, T.D. Kirsch. 2013. The Human Impact of Floods: A Historical Review of Events 1980–2009 and Systematic Literature Review. *PLOS Currents* Disasters, April 16: 5. http://doi.org/10.137/currents.dis.f4deb4579036b07c09daa98ee811a.

Duclos, P.J., L.M. Sanderson, and K.C. Klontz. 1990. Lightning-Related Mortality and Morbidity in Florida. *Public Health Report* 105(3): 276–282.

Eberhart-Phillips, J.E., T.M. Saunders, A.L. Robinson, D.L. Hatch, and R.G. Parrish. 1994. Profile of Mortality from the 1989 Loma Prieta Earthquake Using Coroner and Medical Examiner Reports. *Disasters* 18(2): 160–170.

Edison, M., J.A. Lybargen, J.E. Parsons, J.N. Maccormack, and J.I. Freeman. 1990. Risk Factors for Toronto Injuries. *International Journal of Epidemiology* 19(4): 1051–1056.

Ellidokuz, H., R. Ucku, U.Y. Aydin, and E. Ellidokuz. 2005. Risk Factors for Death and Injuries in Earthquake: Cross-sectional Study from Afyon, Turkey. *Croatian Medical Journal* 46: 613–618.

Elnashai, A.S., B. Gencturk, O.S. Kwon, L. Al-Qadi, Y. Imad, J.R. Hashash, S.J. Roesler, S.H. Kim, J. Jeong, J. Dukes, and A. Valdivia. 2011. *The Maule (Chile) Earthquake of February 27, 2010: Consequence Assessment and Case Studies*. Mid-American Earthquake Center, Report No. 10-04.

Elsom, D.M., and T.D.C. Webb. 2015. Lightning Injuries and Fatalities in the United Kingdom 2014 and a Summary of Personal-Injury Lightning Incidents from 1988 to 2014. *International Journal of Meteorology, United Kingdom* 40(391): 84–91.

Emch, M. E.D. Root, and M. Carrel. 2017. *Health and Medical Geography*. New York, NY: The Guilford Press.

Ensor, B.E., and M.O. Ensor. 2009. Hurricane Mitch: Root Causes and Response to the Disaster. In *The Legacy of Hurricane Mitch: Lessons from Post-Disaster Reconstruction in Honduras*, edited by Ensor, M.O., pp. 22–46. Tucson, AZ: The University of Arizona Press.

FitzGerald, G., W. Du, A. Jamal, M. Clark, and X.Y. Hou. 2010. Flood Fatalities in Contemporary Australia (1997–2008). *Emergency Medical Australia* 22(2): 180–186.

Frank, N.L., and S.A. Husain. 1971. The Deadliest Tropical Cyclone in History. *Bulletin of the American Meteorological Society* 52(6): 438–445.

Franklin, T.L., R.D. Roueche, and D.O. Prevatt. 2015. Comparison of Two Methods of Near-Surface Wind Speed Estimation in the 22 May, 2011 Joplin, Missouri Tornado. *Journal of Wind Engineering and Industrial Aerodynamics* 138: 87–97.

Glass, R.I., J.J. Urritia, S. Sibony, H. Smith, B. Garcia, and L. Rizzo. 1977. Earthquake Injuries Related to Housing in a Guatemalan Village. *Science* 197: 638–643.

Green, R., and S. Miles. 2011. Social Impact of the 12 January 2010 Haiti Earthquake. *Earthquake Spectra* 27(S1): 447–462.

Guha-Sapir, D., F. Vos, R. Below, and R.S. Ponserre. 2012. *Annual Disaster Statistical Review 2011: The Numbers and Trends.* Brussels: Center for Research on the Epidemiology of Disasters.

Guzzetti, F., C.P. Stark, and S.P. Salvati. 2005. The Impact of Landslides in the Umbria Region, Central Italy. *Natural Hazards and Earth Systems Science* 3: 469–486.

Hanna, E.G., T. Kjellstorm, C. Bennett, and K. Dear. 2011. Climate Change and Rising Heat: Population Health Implications for Working People in Australia. *Asia Pacific Journal of Public Health* 23(2): 145–265.

Hofer T, and B. Messerli. 2006. *Floods in Bangladesh: History, Dynamics and Rethinking the Role of the Himalayas.* Tokyo: United Nations University Press.

Holle, R.I. 2016. A Summary of Recent National-Scale Lightning Fatalities Studies (http://doi.org/10.1175/WCAS-D-15-0032.1 – last accessed April 14, 2020).

Howe, N. 2019. More Women Die in Natural Disasters – Why? And What Can Be Done? *Brink*, 25 April (www.brinknews.com/gender-and-disasters/ – last accessed June 13, 2020).

Hu, P., Q. Zhang, P. Shi, B. Chen, and J. Fang. 2018. Flood-Induced Mortality Across the Globe: Spatiotemporal Pattern and Influencing Factors. *Science of the Environment* 643: 171–182.

Ikeda, K. 1995. Gender Differences in Human Loss and Vulnerability in Natural Disasters: A Case Study from Bangladesh. *Indian Journal of Gender Studies* 2(2): 171–193.

Jensen, J.D., and A.L. Vincent. 2019. Lightning Injuries. StatPears (Internet) (www.ncbi.nlm.nih.books/NBK441920/– last accessed April 11, 2020).

Jensenius, Jr., J.S. 2019. *A Detailed Analysis of Lightning Deaths in the United States from 2006 through 2018.* National Lightning Society Council.

Johnston, D., S. Standring, K. Ronan, M. Lindell, T. Wilson, J. Cousins, E. Aldridge, M.W. Ardagh, J.M. Deely, S. Jensen, T. Kirsch, and R. Bissell. 2014. The 2010/2011 Canterbury Earthquakes: Context and Cause of Injury. *Natural Hazards* 73: 627–637.

Jonkman, S.N., and I. Kelman. 2005. An Analysis of the Causes and Circumstances of Flood Disaster Deaths. *Disasters* 29(1): 75–97.

Kahn, M.E. 2003. The Death Toll from Natural Disasters: The Role of Income, *Geography, and Institutions. Mimeo.* Medford, MA: Tuffs University.

Kahn, M.E. 2005. The Death Toll from Natural Disasters: The Role of Income, Geography, and Institutions. *Economics Statistics* 87: 271–284.

Karstens, C.D., W.A. Gallus Jr., B.D. Lee, and C.A. Finley. 2013. Analysis of Tornado-Induced Tree Fall Using Aerial Photography from the Joplin, Missouri, and Tuscaloosa-Birmingham, Alabama Tornado of 2011. *Journal of Applied Meteorology and Climatology* 52: 1049–1068.

Keatinge, W. 2003. Death in Heat Waves: Simple Preventive Measures May Help Reduce Mortality. *BMJ* 327(7414): 512–513.

Keil, R., and H. Ali. 2006. The Avian Flu: Some Lessons Learned from the 2006 SARS Outbreak in Toronto. *Area* 38: 227–239.

Kennedy, I.T.R., D.N. Petley, R. Williams, and V. Murray. 2015. A Systematic Review of the Health Impacts of Mass Earth Movement (Landslides). *PLoS Currents*, April 30, 7. https://doi.org/10.1371/currents.dis.1d49e84c8bbe678b0e70cf7fc35d0b77.

Kenney, W.L., D.H. Craighead, and L.M. Alexander. 2014. Heat Waves, Aging, and Human Cardiovascular Health. *Medical Science Sports Exercise* 46(10): 1891–1899.

Khan, F., and D. Mustafa. 2007. Navigating the Contours of the Pakistani Hazardscapes: Disaster Experience Versus Policy. In *Working with the Winds of Change: Towards Strategies for Responding to the Risk Associated with Change and Other Hazards*, edited by Moench, M., and Dixit, A., pp. 193–234. Kathmandu: Nepal Institute for Social and Environmental Transition (ISET).

Khazai, B., J. Anthorn, T. Girard, S. Brink, J. Daniell, B. Muhr, V. Florchinger, and T. Kunz-Plapp. 2015. *Shelter Response and Vulnerability of Displaced Populations in the April 25, 2015 Nepal Earthquake*. Heidelberg, Germany: South Asia Institute, Heidelberg University.

Krishnaraj, M. 1997. Gender Issues in Disaster Management: The Latur Earthquake. *Gender, Technology and Development* 1(3): 395–411.

Kuligowski, E.D., F.T. Lombardo, L.T. Phan, and M.L. Levitan. 2013. *Technical Investigation of the May 22, 2011 Tornado in Joplin, Missouri*. Washington, DC: National Institute of Standards and Technology (NIST).

Kunii, O., T. Kunori, K. Takahashi, M. Kaneda, and N. Fuke. 1996. Health Impact of 1996 Tornado in Bangladesh. *The Lancet* 348(9029): 757.

Lambie, E., T. Wilson, D.M. Johnston, S. Jensen, E. Brogt, E. Emma, H. Doyle, M.K. Lindell, and W.S. Helton. 2016. Human Behaviour during and Immediately following Earthquake Shaking: Developing a Methodological Approach for Analysing Video Footage. *Natural Hazards* 80(1): 249–283.

Logue, J.N., M.E. Melick, and H. Hansen. 1981. Research Issues and Directions in the Epidemiology of Health Effects of Disasters. *Epidemiologic Review* 3: 140–162.

Madrigano, J., K. Ho, S. Johnson, P.L. Kinney, and T. Matte. 2015. A Case-Only Study of Vulnerability to Heat Wave-Related Mortality in New York City (2000–2011). *Environmental Health Perspectives* 123(7): 672–678.

Mathue-Giangreco, M., W. Mack, H. Seligson, and L.B. Bourque. 2001. Risk Factors Associated with Moderate and Serious Injuries Attributable to the 1994 Northridge Earthquake, Los Angeles, California. *Annals of Epidemiology* 11: 347–357.

Margesson, R., and M. Taft-Morales. 2010. *Haiti Earthquake Crisis and Response*. Washington, DC: Congressional Research Service.

Masai, R., L. Kuzunishi, and T. Kondo. 2009. Women in the Great Hanshin Earthquake. In *Women, Gender and Disaster: Global Issues and Initiatives*, edited by Enarson, E., and P.G.D. Chakrabarti, pp. 131–141. New Delhi: Sage.

Meade, M.S. 1977. Medical Geography as Human Ecology: The Dimension of Population Movement. *Geographical Review* 67: 379–393.

Morrow, B.H. 1999. Identifying and Mapping Community Vulnerability. *Disasters* 23: 1–18.

Naughton, M.P., A. Henderson, M.C. Mirabelli, R. Kaiser, J.L. Wilhelm, S.M. Kieszak, C.H. Robin, and M.A. McGeehin. 2002. Heat-Related Mortality during

a 1999 Heat Wave in Chicago. *American Journal of Preventive Medicine* 22(4): 221–227.
Nishikiori, N., T. Aba, D.G.M. Costa, S.D. Dharmaratne, O. Kuii, and K. Moji. 2006. Who Died as a Result of Tsunami? Risk Factors of Mortality Among Internally Displaced Persons in Sri Lanka: A Retrospective Cohort Analysis. *BMC Public Health* 6: 73.
Noji, E.K., G.D. Kelen, H.K. Armenian, A. Oganessian, N.P. Jones, and T. Sivertson. 1990. The 1988 Earthquake in Soviet Armenia: A Case Study. *Annals of Emergency Medicine* 19(8): 891–897.
Oppenheimer, C. 2003. Climatic, Environmental and Human Consequences of the Largest Known Historic Eruption: Tamboro Volcano (Indonesia) 1815. *Progress in Physical Geography* 27: 230.
Oxfam. 2005. *The Tsunami's Impact on Women.* Oxfam Briefing Note, March.
Padli, J., M.S. Habibullah, and A.H. Baharom. 2013. Determinants of Flood Fatalities: Evidence from a Panel Data of 79 Countries. *Journal of Social Sciences & Humanity* 21: 81–98.
Parrish, H.M., A.S. Barker, and F.M. Bishop. 1964. Epidemiology in Public Health Planning for Natural Disasters. *Public Health Report* 79: 863–867.
Paul, B.K. 2009. Why Relatively Fewer People Died? The Case of Bangladesh's Cyclone Sidr. *Natural Hazards* 50: 289–304.
Paul, B.K. 2011. *Environmental Hazards and Disasters: Contexts, Perspectives and Management.* Hoboken, NJ: Wiley-Blackwell.
Paul, B.K., and S. Dutt. 2010. Hazard Warnings and Responses to Evacuation Orders: The Case of Bangladesh's Cyclone Sidr. *Geographical Review* 100: 336–355.
Paul, B.K., and S. Mahmood. 2016. Selected Physical Parameters as Determinants of Flood Fatalities in Bangladesh, 1972–2013. *Natural Hazards* 83:1703–1715.
Paul, B.K., and A. Ramekar. 2018. Host Characteristics as Risk Factor of the 2015 Earthquake- Induced Injuries in Nepal: A Cross-Sectional Study. *International Journal of Risk Reduction* 27: 118–126.
Paul, B.K., H. Rashid, M. Shahidul, and L.M. Hunt. 2010. Cyclone Evacuation in Bangladesh: Tropical Cyclone Gorky (1991) vs. Sidr (2007). *Environmental Hazards* 9: 89–101.
Paul, B.K., and M. Stimers. 2012. Exploring Probable Reasons for Record Fatalities: The Case of 2011 Joplin, Missouri, Tornado. *Natural Hazards* 64: 1511–1526.
Paul, B.K., and M. Stimers. 2014. Spatial Analyses of the 2011 Joplin Tornado Mortality: Deaths by Interpolated Damage Zones and Location of Victims. *Weather, Climate, and Society* 6(2): 161–174.
Paul, B.K., S. Mahmood, and A. Ramekar. 2018. Analysis of 2015 Nepal Earthquake Mortality Using Spatial Bayesian Model. Presented at the Annual Conference of the Association of American Geographers held in New Orleans, April 6–9.
Pedan, A.E., and R.C. Franklin. 2019. Exploring Flood-Related Unintentional Fatal Drowning of Children and Adolescents Aged 0–19 Years in Australia. *Safety*, 5, 56. https://doi.org/10.3390/safety 5030046.
Peek-Asa, C., J.F. Kraus, L.B. Bourque, D. Vimalachandra, J. Yu, and J. Abrams. 1998. Fatal and Hospitalized Injuries Resulting from the 1994 Northridge Earthquake. *International Journal of Epidemiology* 27: 459–465.
Peek-Asa, C., M.R. Ramirez, K. Shoaf, H. Seligson, and J.F. Kraus. 2000. GIS Mapping of Earthquake-Related Deaths and Hospital Admissions from the 1994 Northridge, California, Earthquake. *Annals of Epidemiology* 1(1): 5–13.

Peek-Asa, C., M.R. Ramirez, H. Selingson, and K. Shoaf. 2003. Seismic, Structural, and Individual Factors Associated with Earthquake Related Injury. *Injury Prevention* 9: 62–66.

Petal, M. 2011. Earthquake Casualties Research and Public Education. In *Human Casualties in Earthquake: Advances in Natural and Technological Hazard Research, 29*, edited by Spence, R., S. Emily, and S. Charles, pp. 25–50. New York, NY: Springer.

Pradhan, E.K., K.P. West, P.H.J. Katz, S.C. LeClerq, S.K. Khatry, and S. Ram. 2007. Risk of Flood-Related Mortality in Nepal. *Disasters* 31:57–70.

Provic, K.A. 2012. Terrain and Land Cover Effects of the Southern Appalachian Mountains on the Rotational Low-Level Wind Fields of Super Cell Thunderstorms. *M.A. Thesis*, Virginia Polytechnic Institute and State University. Blacksburg, VA.

Rahman, M.K., B.K. Paul, A. Custis, and T.W. Scmidlin. 2015. Linking Coastal Disasters and Migration: A Case Study of Kutubdia Island, Bangladesh. *The Professional Geographer* 67(2): 218–228.

Ramirez, M., and C. Peek-Asa. 2005. Epidemiology of Traumatic Injuries from Earthquake. *Epidemiologic Review* 27(1): 47–55.

Riley, M. n.d. Types of Volcanic Hazards (www.geo.mtu.edu/volcanoes/hazards/primer/ – last accessed April 15, 2020).

Rofi, A., S. Doocy, and C. Robinson. 2006. Tsunami Mortality and Displacement in Aceh Province, Indonesia. *Disaster* 30(3): 340–350.

Roy, N., H. Shah, V. Patel, and R.R. Caughlin. 2002. The Gujarat Earthquake (2001) Experience in a Seismically Unprepared Area: Community Hospital Medical Response. *Prehospital Disaster Medicine* 17(4): 186–195.

Sanchez, C., T.-S., Lee, S. Young, D. Batts, J. Benjamin, and J. Malilay. 2009. Risk Factors for Mortality during the 2002 Landslides in Chuuk, Federal States of Micronesia. *Disasters* 33(4): 705–720.

Schmidlin, T.W., and P.S. King. 1995. Risk Factors for Deaths in the 27 March 1994 Georgia and Alabama Tornadoes. *Disasters* 17(1): 170–177.

Seligson, H., and K. Shoaf. 2003. Human Impacts of Earthquakes. In *Earthquake Engineering Handbook*, edited by Chen, W.F., and C. Seawhorn. Boca Raton, Florida: CRC Press.

Sen, A. 1981. Ingredient of Famine Analysis: Availability and Entitlements. *Quarterly Journal of Economics* 96: 433–464.

Seo, S.N. 2015. Fatalities of Neglect: Adapt to More Intense Hurricanes Under Global Warming? *International Journal of Climatology* 35(2): 3505–3514.

Shapira, S., L. Aharonson-Daniel, I.M. Shohet, C. Peek-Asa, and Y. Bar-Dayan. 2015. Integrating Epidemiological and Engineering Approaches in the Assessment of Human Casualties in Earthquakes. *Natural Hazards* 78(2): 1447–1462.

Sharma, S. 1995. Drought, Mortality and Social Structure. *Environmental Education Information* 14(1): 85–94.

Simmons, K.M., and D. Sutter. 2011. *Economic and Societal Impacts of Tornadoes.* Boston, MA: American Meteorological Society.

Simmons, K.M., and D. Sutter. 2012. *Deadly Season: Analysis of the 2011 Tornado Outbreaks.* Boston, MA: American Meteorological Society.

Smith, T. 1934. *Parasitism and Disease.* New York, NY: Hafner Publishing Company.

Stanke, C., M. Kerac, C. Prudhomme, J. Medlock, and V. Murray. 2013. Health Effects of Drought: A Systematic Review of Evidence. *PLoS Current*, June 5. https://doi.org/10.1371/currents.dis.7a2cee9e980191ad7697b570bee46004.

Stimers, M.J., and B.K. Paul. 2016. Toward Development of the Tornado Impact-Community Vulnerability Index. *Journal of Geography and Natural Disasters* 6: 1–11.

Stimers, M.J., and B.K. Paul. 2017. Can Elevation be Associated with the 2011 Joplin, Missouri, Tornado Fatalities? An Empirical Study. *Journal of Geography and Natural Disasters* 7(3): 1–7.

Stocker, T.F. 2014. *Climate Change 2013: The Physical Science Basis: Working Group 1 Contribution in the Fifth Assessment Report of the Intergovernmental Panel on Climate Change.* Cambridge: Cambridge University Press.

Sugimoto, T.D., A.B. Labrique, S. Ahmad, M. Rashid, A.A. Shamim, B. Ullah, R.D. Klemm, P. Christian, and K.P. West Jr. 2011. Epidemiology of Tornado Destruction in Rural Northern Bangladesh: Risk Factors for Death and Injury. *Disasters* 35(2): 329–345.

Sullivan, K., and S. Hossain. 2009. Earthquake Mortality in Pakistan. *Disasters* 34(1): 176–183.

Tantiworawit, P., C. Yuphan, T. Wangteerapraprasert, E. Yodkaiw, I. Ieowongjaroen, S. Iamsirithaworn, and R. Bauthong. 2012. Risk Factors of Flood-Related Mortality in Phichit Province, *Thailand. OSIS* 9(4): 11–17.

Toya, H., and M. Skidmore. 2006. Economic Development and the Impacts of Natural Disasters. *Economic Letter* 94: 20–25.

Vandentorren, S., P. Bretin, A. Zeghnoun, L. Mandereau-Bruno, A. Croisier, C. Cochet, J. Riberon, I. Siberan, B. Declereq, and M. Ledrans. 2006. August 2003 Heat Wave in France: Risk Factors for Death of Elderly People Living at Home. *European Journal of Public Health* 16(6): 830–833.

Walsh, B. 2010. After the Destruction: What will it Take to Rebuild Haiti. *Time* 16 January.

Walker, G.P.L. 1983. Ignimbrite Type and Ignimbrite Problems. *Journal of Geothermal Research* 17: 65–88.

Wurman, J., C. Alexander, P. Robinson, and Y. Richardson. 2007. Low-Level Winds in Tornadoes and Potential Catastrophic Tornado Impacts in Urban Areas. *Bulletin of the American Meteorological Society* 88: 31–46.

Yamada, S., R.P. Gunatilake, T.M. Roytman, T. Fernando, and L. Fernando. 2006. The Sri Lanka Tsunami Experience. *Disaster Management and Response* 4(2): 38–48.

Yin, Q., and J.F. Wang. 2017. The Association between Consecutive Days' Heat Wave and Cardiovascular Disease Mortality in Beijing, China. *BMC Public Health* 17.

6 Conclusion

This chapter provides a summary of the major findings of this book. It suggests how to reduce deaths from different types of natural disasters by analyzing what to do or not to do before, during, and after such extreme events. These suggestions can be considered recommendations to assist in drastically reducing mortality from natural disasters. Note that this reduction can be achieved through sincere initiatives and efforts of both the respective governments and at-risk populations of disaster-prone areas. The involvement of one without the other will not help reach the desired goal of losing fewer lives to disaster events. For example, if a disaster forecast is issued but does not reach to the population at risk, then the warning system is not sufficient. Conversely, if such a population receives the warning, but ignores it totally, then the forecast system is useless. Finally, several future research directions are suggested.

Note that there are many commonalities across all disasters, as well as a specific way to reduce mortality from a specific disaster. For example, risks of deaths can be reduced by improving forecasting, early warning dissemination systems, and monitoring of disasters in all countries of the world. Adequate and timely warning by the appropriate public authorities is the best mitigation measure against deaths result from natural disasters. The disaster forecasting system must be provided sufficient lead time for at-risk population to respond appropriately.

Increasing lead time increases the potential to reduce casualties. The exception to this is found in tornado events where longer lead time leads to delay in taking safe shelter. However, the forecast must be sufficiently accurate and credible to promote confidence so that people will respond when warned. If forecasts are repeatedly inaccurate, they will introduce false alarms (also known as "crying wolves"), which may challenge the future warning credibility, eventually reducing compliance with evacuation order (Dow and Cutter 2000). At the same time, the authorities should discourage a "shadow evacuation."[1]

Similarly, increasing awareness and development of appropriate management plans are also useful in developing and implementing sound policies to prevent or reduce disaster-attributed mortality. In this regard, public restrictions on the development of structures in potential hazard zones, including urban areas, will help reduce the suffering and death of people in hazard-prone areas. Along with these, individual and household preparedness and compliance with safety tips before, during, and after a disaster will also be useful in decreasing the death toll. Exposure from all extreme events can further be reduced if the appropriate authorities initiate evacuation of people situated in harm's way, clearing them of the impeding hazards (this is particularly for tropical cyclone events). At the same time, people should make all efforts to fully comply with all evacuation orders (Dow and Cutter 2000; Paul and Dutt 2010; Paul et al. 2010).

In a considerable number of cases, conservative and illiterate people exhibit a kind of fatalism accompanied by a pessimistic attitude that leads them from not to evacuate, choosing to take refuge in cyclone shelters instead. Similarly, women are found to have felt excluded from cyclone shelters that are not sensitive to their needs and typically dominated by men. In many developing countries, there is an inadequate number of cyclone shelters, and therefore there is an urgent need to construct additional shelters as well as improve existing ones.

Greater inclusion and participation will lead to disaster risk prevention; this means that both males and females should actively participate in the decision-making process for any measure to reduce disaster deaths. To foster their self-efficacy, they should have easy access to available resources, acquire skills, and build their capacity to lower such deaths. Also, attempts at lowering the death toll can benefit by intensifying disaster risk communication, which has two purposes: (1) sharing the knowledge about the nature and magnitude of risk and (2) taking risk-reducing actions. Risk communication also empowers, motivates, and encourages people to be proactive rather than passive about reducing the risk of deaths from natural disasters. Compliance with evacuation can be improved by educating public regarding the advantages of such an action. Further, fatalism can be removed from the minds of some people through risk communication and education.

Awareness through public education is also useful in decreasing the loss of life due to natural disasters. Studies reviewed in this book suggest that people, and even some personnel involved in disaster preparedness, response, and management, are not fully aware of some of the aspects of a particular extreme event. Knowledge about these aspects can raise awareness as well as can save many lives. For example, people in developed and developing countries generally seek shelter under isolated trees or tall objects to avoid rain, but fail to understand that such trees and objects do not protect from lightning (Duclos et al. 1990).

Public disaster education should make it clear that tsunami waves can travel to the coast multiple times with a temporal gap of 10–30 minutes. Moreover, wave activities often last for several hours. Many people believe that a tsunami wave travels only once on the coast and then moves inland. The consequences of these assumptions are deadly. People also should know that deaths can be caused when tsunami waves recede. Similarly, cyclone-induced storm surges hit coastal areas more than one time. Also, one should not leave the house while tropical cyclones and tornadoes are occurring. Through public education, people should know what to do before, during, and after each type of natural disaster.

One concern in studying disaster-induced deaths is the quality of mortality data. This is for two reasons: there are many sources of disaster death data, and there is a lack of agreement among them. For example, there are at least four official sources of disaster mortality data in the United States: (1) public health and vital statistics departments, (2) the Federal Emergency Management Agency (FEMA) funeral claims database, (3) the American Red Cross' (ARC) mortality surveillance system, and (4) the National Oceanic and Atmospheric Administration (NOAA) and National Weather Service (NWS) storm database. In addition, there are other nonofficial sources of disaster mortality data such as the Spatial Hazard Event and Loss Database for the United States (SHELDUS), developed by the Hazards and Vulnerability Research Institute at the University of South Carolina, Columbia, South Carolina. However, the Centers for Disease Control and Prevention (CDC) found considerable differences between the final number of deaths recorded by the different official sources for the same disasters when declared a federal disaster (CDC 2017). Therefore, the selection of data source is crucial for investigating different aspects of disaster mortality.

Summary

Chapter 1 has begun by providing salient information on disasters and deaths, along with objectives of the book, chapter arrangement, and the need for the book. A large part of Chapter 1 was devoted to discussing types of and problems with disaster deaths data. While direct death data are relatively easy to collect and understand, indirect deaths are frequently challenging to identify and take time even years after the disasters. At national and subnational levels, many organizations are responsible to collecting disaster deaths data; the number of deaths differs among sources as they collect and compile data in a variety of ways without following systematic and standardized method. The reliability and quality of data are often compromised because they are collected data rapidly to use in short-term disaster response efforts.

Complete and reliable disaster mortality data can be collected through by introducing civil and vital registration systems in all countries; these data can be used for long-term mitigation purposes. Although there are problems with international organizations, such as the CRED and the DesInventar, these organizations collect disaster deaths and other relevant data using standardized methods. Because of compiling these data is delayed, they can be used for long-term mitigation measures to reduce disaster-induced deaths. Irrespective of the sources of data, disaster deaths data should be sufficiently consistent with each other and comparable across any geographic scale.

Chapter 2 examined the reasons for unexpected death toll numbers caused by 15 natural disasters. Nine of these extreme events experienced far more deaths than expected, while the remaining six resulted in far fewer than expected. In both cases, one or more physical characteristics of disasters either favorably or adversely worked to affect the number of deaths. For example, the 2010 Haiti earthquake killed a record number of people primarily due to four reasons: (1) depth of epicenter, (2) epicentral distance to densely populated areas, (3) massive size of vulnerable population, and (4) lack of public preparedness. The first two reasons were associated with two critical physical dimensions of the earthquake. The Haiti earthquake was not considered a "high" magnitude event, but it originated at a shallow depth, and its epicenter was close to the capital city of the country. Both are leading risk factors for earthquake-induced deaths. Two other factors were added to trigger a high death toll. The Haiti government was not prepared for the event because it was not considered an earthquake-prone country. For this reason, and combined with widespread poverty, Haitians did not need to build houses according to seismic codes. Thus, multiple factors interacted with each other to produce unexpected death toll numbers.

Alone, or as a combination of two or more factors such as lack of early warnings, high incidence of false warnings, and an inadequate number of public shelters were directly associated with too many deaths caused by flood, cyclone, and tornado events as discussed in Chapter 2. Early warnings, faith in such warnings, and a satisfactory rate of compliance with evacuation order were associated with few deaths resulting from five events included in this chapter. Compliance with evacuation orders and an inadequate number and quality of public shelters still act as barriers for not fully utilizing these facilities. For example, a further increase in compliance with evacuation orders is possible by placing cattle and human shelters close to each other. Similarly, an adequate number of public shelters with provision for separate latrines and rooms for males and females will increase in the use of these facilities. Studies (e.g., Paul and Dutt 2010; Ahsan et al. 2016) show that coastal residents in Bangladesh prefer public shelters to be located at one mile apart, as they are not willing to walk more than one mile in severe weather. Thus, they prefer dense small-size public cyclone shelters than dispersed large ones.

The first part of Chapter 3 analyzed 20 disaster-induced death trends for the 25-year study period (1991–2015). Only seven of those 20 examinations showed an upward trend during the period, while the remaining 13 tended downward. All trend analyses were conducted at scales larger than country, which implies that at the country-level trends could be different than the reported trend. However, better preparedness and response contributed to the declining trend in disaster deaths. It was also largely explained by the implementation of mitigation measures to reduce deaths from different natural disasters. The second part of Chapter 3 covered two important topics: one is related to the single most common disaster myth that "disasters cause epidemics," while the second topic is the burial or cremation of bodies. Proper disposal of bodies is vital to kin members for their mental health – they sincerely desire for their loved ones to be buried or cremated according to appropriate local rituals and religious traditions.

Although circumstances and causes, as discussed in Chapter 4, are disaster-specific (especially "causes"), surprisingly, there are common causes among extreme events, even those as apparently divergent as droughts and volcanic eruptions. The root cause of deaths due to those two extreme events is starvation. Similarly, floods and storm surges account nearly two-thirds of deaths resulting from drowning. Collapse of all types of buildings is the root cause of deaths caused by earthquakes and tornadoes. Automobile accidents are the principal cause of deaths for ice and snowstorms. Hyperthermia is the leading cause of deaths stemming from heatwaves.

An understanding of other minor causes is also important from the perspective of the design and to implement of measures to reduce deaths from natural disasters. For example, other causes of flood deaths include being struck by debris in water, vehicle crashes, and the collapse of buildings due to floodwaters. Similarly, minor causes of deaths related to cyclones include poisoning-induced by carbon monoxide, heart attack, hyperthermia, asphyxiation, electrocution, and burns. Note that poisoning typically results from people leaving their generators running indoors. Knowledge about both major and minor causes of disaster-induced deaths is essential because many of these fatalities are easily preventable.

Materials presented in Chapter 5 revealed that both hazards and vulnerable factors appear as essential determinants of deaths resulting from natural disasters. Each of these two broad factors plays different roles in different disasters. For example, physical dimensions such as magnitude, duration, spatial extent, frequency, and temporal spacing are key risk factors of deaths caused by earthquakes. However, only the first three physical characteristics are considered risk factors for deaths induced by droughts. For lightning, only diurnal and seasonal factors determine the extent of deaths. People have little to no control over these characteristics of natural disasters. Although there are some interactions between the two broad types of factors, programs should aim to improve the living conditions of vulnerable populations (e.g., elderly, children, the poor, women, disability, and minority) of a society so they can be well prepared for and adapt to the threat of natural disasters.

Disaster-specific tips and recommendations

Several tips and recommendations are common to all natural disasters. Each household should have emergency plans for different extreme events and occasionally review them with family members. Others include having emergency kits at home and in their car, and staying informed through radio, television, phone, or social media regarding disaster watches and warnings before, during, and after the event. Other tips are disaster-specific.

Earthquake

There are many measures one can entail to reduce their risk of deaths from an earthquake. Long before the tremblor, households should have a plan, which includes exit route from the building and a location at which to reunite after the event. In earthquake-prone regions, each household should keep a fire extinguisher, first aid kit, flashlight, and extra batteries readily available. A fire extinguisher is essential because earthquakes are frequently cause fires, both large and small, which cause a considerable number of deaths. For example, most of the 142,800 deaths caused by the 1923 Great Kanto Earthquake in Japan were due to fire that erupted after the earthquake. At the time of earthquake, many people in the affected areas were preparing a meal in cooking stoves (Schencking 2013). Each household member should learn how to shut off gas, electricity, and water in case lines are damaged. Also, secure heavy furniture and appliances to the walls or floors and keep flammable liquids in closed cabinets.

If household members stay indoors during an earthquake, they should take cover under heavy furniture (e.g., beds, tables, or desks), cover heads with clothes to protect themselves from falling objects, and avoid windows and outside doors. If a person is outdoors, he/she should stay in the open areas away from powerlines, buildings, or anything that might fall. For the duration of an earthquake, it is advised not to use elevators as they may be struck. If one is in a car, he or she needs to stop the car and stay inside until the earthquake subsides. Do not stop cars under bridges or overpasses, which may collapse due to the shaking caused by earthquake-induced energy moving through the ground.

After the earthquake, people stay out of damaged buildings and check utilities. If there is damage, turn the utility off. Never use matches, candles, and lighters inside. To avoid injury from broken glasses and debris, use sturdy shoes and appropriate clothing. Be prepared for aftershocks and avoid going to beaches as tsunamis waves sometimes strike after an earthquake – tsunami waves can cause deaths from drowning.

As indicated, nearly three-quarters of all deaths in earthquakes are caused by building collapse (Cross 2015). Low cost and poorly built buildings are most likely to fail. Thus, the most effective way to save many lives from an earthquake is to enforce the latest building codes for new buildings and retrofit existing ones. In developing countries, codes are not always strictly enforced; therefore, in these countries, the respective government should monitor the enforcement of building codes. Due to widespread poverty, applying of building codes requires financial support from the government and international agencies.

Droughts

Starvation is the root cause of death ascribed to drought disasters. The event reduces crop yields, causing a shortage while also increasing food prices, potentially leading to malnutrition. Therefore, there is a need to establish food subsidy programs for people in drought-affected areas. The national government and international agencies such as World Food Program (WFP) should initiate food distribution to the poor, either free or at low cost. The WFP is the leading humanitarian organization in the delivery of food assistance in emergencies and working with communities to improve nutrition. It purchases and stores food in its warehouses scattered all over the developing world. During droughts that may last for months or years, the government should stop the hoarding of food by individuals and business people.

However, drought increases the risk of death through valley fever, which is transmitted to humans when spores in the soil become airborne and are inhaled. It triggers secondary disasters, such as wild or forest fires and dust storms. These two secondary disasters can reduce air quality and hence affect people with already compromised health. The low quality of air causes chronic respiratory diseases and intensifies deaths from these illnesses. Poor air quality can also increase the spread of infectious diseases such as gastrointestinal illnesses. Note that people in developed countries also die from driving car in dust storms due to reduced visibility.

To conserve the water, people can install low-flow faucet aerators to reduce water use while maintaining proper hygiene. Restrictions of used of water for agricultural production often lead to increased depression, and anxiety, which, in turn, increases suicide rate among farmers. Drought not only reduces water quality but can lower the water level in bodies of water, and cause water to stagnant, providing ideal breeding grounds for mosquitoes, which in turn can cause an increase of malaria incidence during the drought, typically resulting in the deaths of significant numbers of children and the elderly.

Floods

Drowning is the leading cause of death during floods, both in developed and developing countries. Developing countries suffer the highest death tolls from the occurrence of flooding, but the reasons for drowning also differ from those of developed countries. Almost all flood-related drowning deaths are preventable by taking simple precautions such as close supervision of children by an adult family member, avoiding flooded areas and floodwater, not allowing children to play in floodwater areas, and evacuating areas that are subject to flooding (e.g., moving to higher ground or otherwise away from the flooded area). Another way is to teach children and adults swim – all schools in developing countries should introduce swimming lessons in their curricula. Also, managing flood risks through better flood preparedness can reduce drowning deaths during flooding.

In both developed and developing countries, flood deaths are associated with an increased risk of infection due to contamination of sources of drinking water sources. Also, floodwaters tainted with sewage may contain *Escherichia coli* (*E. coli*) bacteria or other hazardous substances. As the Hyogo Framework recommended, one way to minimize deaths from waterborne diseases is to use bottled or treated water for drinking and cooking purposes (Green et al. 2019). People should not bathe in water that may be contaminated with sewage or toxic chemicals. Ideally, exposure to floodwater should be avoided altogether, as it may contain sharp objects such as broken glass or metal fragments that can cause injury and lead to a fatal infection.

In developed countries, flood deaths can be dramatically reduced by avoiding driving through flooded areas and standing water. As little as six inches (15.24 cm) of water can cause a driver to lose control of a vehicle. In these countries, a reasonable number of deaths attributed to floods are caused by carbon monoxide poisoning. These deaths can be prevented by using generators at least 20 feet (6.1 m) from doors, windows, and vents. Alternately, carbon monoxide detectors can be installed and should be routinely checked to ensure proper functioning. During and after flood, electrocution can occur when people make contact with power supplies. After a flood, one should be careful to avoid chemical hazards inside or outside the home and should shut off the electrical power and natural gas or propane tanks in home to avoid fire, electrocution, or explosions.

Tropical cyclones and storm surges

Deaths resulting from tropical cyclones, hurricanes, or typhoons differ significantly between developed and developing countries. Globally, over 90 percent cyclone-related deaths occur in developing countries. Therefore, more attention should be directed toward how to reduce cyclone deaths in these countries. Despite improvements in early warning systems, pre-cyclone evacuation remains a challenge in many developing countries such as Bangladesh, Cuba, India, and the Philippines. Increasing evacuation rate is the key to reduce mortality from a cyclone and associated storm surges in these countries. Thus, only one pre-event activity will reduce many cyclone-related deaths.

Coastal residents of cyclone-prone developing countries should go to nearby cyclone shelters immediately after an evacuation order is issued. Waiting for last moment to comply with such an order, or adopting a "wait-and-see approach," can be deadly. Studies (e.g., Paul and Dutt 2010; Haque et al. 2012; Doocy et al. 2013; Ching et al. 2015) have reported that people died *en route* to shelter because, by the time they decided to evacuate, the cyclone had already made landfall and moved to inland areas. People also died on their way home after they had turned back because many of the shelters were already overcrowded. Another reason to evacuate early is that the intended shelter will still have room to accept incoming evacuees. All household members should evacuate together, although a tendency is observed in some developing countries, such as Bangladesh, where select family members proceed to the shelter while at least one member stays at home to protect household goods. Previous false warnings also limit people's willingness to evacuate to shelters (Haque et al. 2012). Therefore, a reduction in the number of false alarms can reduce death from cyclones.

Evacuation intent is hindered by a host of other factors such as lack of a nearby cyclone shelter or perception of unsafe conditions in the shelter. For example, many coastal residents in Bangladesh suspect that the physical condition of such shelters will not withstand the impending cyclone (Paul and Dutt 2010).[2] Moreover, many of these shelters lack facilities such as latrines and lights; this is a particularly serious concern for women in Bangladeshi culture, as putting men and women who do not know each other together under the same roof is inappropriate (Ikeda 1995; Ahsan et al. 2016). Separate rooms for males and females and the addition of latrines and light to existing public shelters will reduce mortality rate by increasing evacuation rate among both males and females.

Some structural changes in the coastal areas will also reduce mortality not only from cyclones and associated storm surges but also from tsunami waves. The building codes in such areas should be changed to ensure that homes are strong enough to withstand the high winds of cyclones and speed of rushing ocean water. Introducing steel-framed or concrete houses with open ground levels will reduce the mortality rate. After the 2004 Indian Ocean Tsunami, the Sri Lankan Government encouraged coastal residents to rebuild destroyed homes with an open ground level to allow tsunami waters to flow under the structure (Hettiarachchi and Dias 2013). Generally, though, coastal residents cannot afford to incorporate recommended new housing features without help from national or foreign governments, international agencies, and nongovernmental organizations (NGOs) that provide grants for such constriction. Additionally, all homes in coastal areas should be surrounded by trees, which reduce cyclones' wind speed.

After the Cyclone Sidr in 2007, the Bangladesh government decided to build cyclone-resilient core shelters for nearly 79,000 households. The government planned to introduce the Philippines-style Core Shelter Program to coastal areas to improve construction quality of Sidr-induced damaged houses.[3] Unfortunately, the construction of these core shelters has been slow and wildly unsuccessful (Nadiruzzaman and Paul 2013; Paul and Rahman 2013). Along with the Bangladesh government, other foreign countries (e.g., India, Kuwait, and Saudi Arabia) and several NGOs (e.g., Concern Worldwide, Habitat for Humanity, Muslim Aid, World Vision, and the Bangladesh Rural Advancement Committee – BRAC) also committed to building over 100,000 cyclone-resilient core homes in the most severely impacted coastal districts (Paul and Rahman 2013). Because most cyclone deaths are caused by drowning in the storm surges, the construction of coastal embankments or sea walls and afforestation of mangrove trees in the shorelines will be beneficial in decreasing death tolls from cyclones. Mangrove forest can reduce cyclone's wind speed and hence its destructive powers. At the same time, existing coastal embankments should be maintained and repaired as needed (Haque et al. 2012).

Even in developed countries, disaster evacuees are hesitant to leave behind their belongings (Alexander 1993; van Duin and Bezuyen 2000). Many residents in New Orleans, Louisiana, did not evacuate during Hurricane Katrina because many of them were reluctant to leave home without their pets (Cole and Fellows 2008; Dow and Cutter 2000). At that time, there was no shelter for pets.[4] Similarly, in coastal Bangladesh, one of the reasons for not complying with cyclone evacuation orders is that residents who owned cattle were reluctant to evacuate to a cyclone shelter until they could move their stock to a safe location (Haque and Blair 1992: Paul and Dutt 2010). There are separate shelters for cattle, locally called *killas*; however, public cyclone shelters and killas are not located in the same compound. People find it awkward and time-consuming to keep livestock at one site then move to a different site for their own protection.

Adequate preparation helps to reduce cyclone-induced deaths in both developed and developing countries. To withstand the strong wind force of an advancing cyclone, people in developing countries add additional poles to strengthen their houses or tie their houses to nearby matured trees with strong ropes (Paul and Dutt 2010). In developed countries, people generally protect their home's windows and doors with storm shutters or plywood. They also reinforce the garage door and install straps or additional clips to fasten the roof to the home's frame securely. People should leave early enough to avoid being trapped by severe weather if an evacuation order is issued. Before evacuation, they should unplug most of electric appliances. Whether they stay or evacuate, they should trim or remove trees that could fall on home. If they decide not to evacuate, they should stay indoors and away from windows and glass doors.

Tornadoes

Advanced tornado technology could reduce deaths from this type of event. For example, early warning systems have improved remarkably in recent decades, but tornado defenses could still be improved, which would result in fewer deaths. In the United States in the 1980s, forecasters were able to provide about five minutes of advanced warning before a tornado hit a location. In 2020, the average warning time for tornadoes has increased to 15 minutes, providing people in tornado-prone areas more time to seek shelter. Most tornado fatalities are caused by falling and flying debris. During a tornado, there is no completely safe place, but some places are safer than others. When a tornado siren sounds, the CDC in the United States recommends that people who are indoors should not go outside to confirm whether or not the tornado can be seen. Instead, they should immediately go to the basement or an inside room without windows (e.g., interior bathroom, stairwell, closet, or hallway) on the lowest floor. For added protection, people should get under something sturdy (e.g., a table, other furniture, or a bed) and cover their body with a blanket, sleeping bag, mattress, and don a helmet (e.g., bicycle or football) if one is available helmet.

The CDC also recommends avoiding mobile or manufactured homes during a tornado. Because of weak construction and lack of anchoring, these homes are much easier for a tornado to damage and destroy than well-built permanent homes (e.g., wood-framed homes fastened securely to foundations, brick homes, or poured concrete wall homes). Mobile or manufactured homes are often built with lighter materials, which is the major contributor to their affordability. However, this weaker construction does not hold up well to high winds, which can easily flip a mobile home or lift the roof (and subsequently, blow over the walls) of a manufactured home. In the United States, mobile homes account for approximately 44 percent of all tornado deaths.[5] If outdoors, people should seek shelter nearby sturdy buildings. If a person is in a vehicle, they should not attempt to outrun a tornado, but if possible, find shelter or drive at a 90-degree angle away from the tornado. If a sturdy building is not available, do not seek shelter under a highway overpass; alternatively, people should find a ditch or depression in the ground and lie flat with their head covered. People should be aware of the potential for a flash flood event during a tornado as well (Plumer 2014).

Building sturdier homes, schools, and shelters (including storm cellars and safe rooms) in tornado-prone areas will further reduce deaths from this extreme event. Sturdier homes are needed, particularly in states like Oklahoma, where underground shelters are not always practical due to high water table, making basements difficult and costly to construct. Studier homes can withstand heavy winds, and building more durable exterior walls and reinforcing garage doors will reduce tornado fatalities (Plumer 2014). Dome-shaped buildings, such as the ones constructed after the 2007 Greenburg, Kansas, EF-5 tornado, can help withstand stronger winds (Paul and Che 2011). Generally, storm cellars are preferable to a basement, unless the basement contains a safe room. Above-ground safe rooms are less than ideal. Additional costs are needed to construct sturdy buildings, safe rooms, or storm cellars. People in tornado-prone areas often cannot afford the additional cost, and therefore the government should subsidize building these structures. It is also necessary to invest in tornado prevention, particularly for high-risk groups such as the less affluent. Additionally, communal tornado shelters, particularly in medium and large communities, would be useful in helping to reduce tornado fatalities. However, such structures rarely exist within communities in the United States. Regular tornado drills, along with preparedness education, are necessary to reduce fatalities from tornado events.

Heatwaves

An effective way to prevent heatwaves mortality is to use air conditioning, which uses a large amount of energy. Its widespread use not only accelerates global warming, but many people, particularly in developing countries, cannot afford its cost. Heatwaves are predictable, sometimes days and weeks in advance; therefore, steps can be taken to save lives before such waves strike. Since the elderly, the poor, the homeless, and people with preexisting health problems are the most vulnerable groups from these extreme events, they should be protected by proving fans, air conditioning, and water during the heatwaves period. People who work outside are also vulnerable to heatwaves. Note that some agencies can provide air conditioners to those who cannot afford them or subsidies for the purchase of a unit. Some groups may provide support for paying utility bills.

Another option is to open cooling centers with reliable electricity for those without access to air conditioning. The size of the vulnerable population should determine the number of such centers. For a large city, the cooling centers should be distributed across the city. Church and other groups in a community or a city may provide transport costs to reach such centers. Heatwaves deaths can also be reduced by keeping the home cool, avoiding going outside, staying in the shade, keeping the body cool and hydrated, and taking cool showers or baths (see Matthies et al. 2008).

Ice and snow storms

As noted in Chapter 4, Mersereau (2013) identified seven common causes of death during winter ice and snow storms: (1) car accidents, (2) slip and falls, (3) snow shoveling, (4) hypothermia, (5) falling through thin ice and into a water body, (6) falling trees and ice, and (7) roof and building collapse due to weight of ice and snow. People often die from a heart attack during snow shoveling due to overexertion. Although not mentioned by Mersereau (2013), blizzard and other winter storms also kill people because of carbon monoxide poisoning. The best way to reduce deaths from these events is to stay inside and dress warmly if one must venture outside. If outside and trapped in a car during a snowstorm, the best thing to do is turn it off and remain in the vehicle. Turn it on every once in a while, to heat it, and then turn it back off to conserve gas. Continue to check the tailpipe each time the heater is turned on. It is advisable to watch for signs of hypothermia (e.g., memory loss, disorientation, slurred speech, and drowsiness) as well as frostbite, which can be recognized by loss of feeling in and discoloration of extremities (usually redness at the onset), and hard and pale skin in the later stage.

During blizzards and other winter storms, power outages are widespread. Thus, a generator can provide necessary power during an outage, but caution should be exercised to avoid carbon monoxide poisoning. In this context, battery-operated carbon monoxide detectors are valuable items for each household. Roof collapse can be presented by making it stronger before the winter season. Other helpful tips are to remove dead or rotting trees and branches around the home that could fall and damage the structure or injury or death to people. Clear clogged rain gutters to allow water to flow away from home, as melting snow and ice can build up if gutters are clogged with debris. Like cooling centers, after a winter storm warming centers are opened, if there is a need, take shelter in such centers.

Volcanic eruptions

Unlike many natural disasters, the prediction of imminent volcanic eruptions is possible through monitoring volcanoes as there are enough warning signs for such eruptions. These include changes in magnetic fields around a volcano, ground deformation, changes in the groundwater system, and changes in heat flow and gas composition (Brown et al. 2017). If an authority issues a volcanic eruption evacuation order, people in the evacuation zone should follow the order immediately and flee from the volcano area to avoid flying debris, lava flows, hot gases, lahars, and lateral blasts. People should also avoid areas downstream of the eruption, and all low-lying places as lava and mudflows are likely to follow into those areas.

Before crossing a bridge in the evacuation zone, people should look upstream and not cross the bridge if a lahar is approaching. People need to protect themselves from falling ash and cover their nose and mouth with a mask or clean cloth to avoid breathing in ashes and wear protective glasses to shield their eyes. One should not drive in heavy ash fall. Cover water containers and food to avoid contamination with ash. After a volcanic eruption, people should stay indoors and away from volcanic ash fall, and while in a house, close all doors and windows to avoid lettering in ashes.

Other natural disasters

Lightning, landslides, and avalanche are covered under this sub-heading. Protection from these three natural disasters means preventing deaths caused by these events. Lightning always accompanies thunderstorms. If the weather forecast calls for thunderstorms, the best suggestions are to go indoors, avoid concrete floors or walls, not to take a shower, or touch any metal objects, including electric equipment of all types, as lightning can travel through the paths provided by those items. People should stay off corded phones and away from doors, windows, and porches. If outdoors, do not take shelter under the tallest object (e.g., under a tree). Lightning frequently strikes trees, and it quickly travels through them. Stay away from power lines and find a safe, enclosed shelter. People should avoid any water sources such as ponds, lakes, and rivers.

Lowering avalanche deaths can possible by improving avalanche forecasting and focused education tailored to at-risk groups. Safe travel techniques (e.g., have an escape route planned and using slope cuts) are also useful. In the context of landslides, deaths induced by these events can be prevented by modifying slope geometry, constructing piles and retaining walls and other structures, diverting debris pathways, and rerouting surface and subsurface drainage. Other helpful ways to reduce landslide deaths are restricting certain types of land use where slope stability is in question, installing early warning systems, preserving vegetation, protecting rock fall, and understanding landslide warning signs. Another effective way is not to live or work in areas that have a history of landslides.

Future research

Worldwide death analyses have received much less attention for some natural disasters (e.g., tornadoes, tsunami, blizzards, and heatwaves). Thus, more studies are needed in these areas of research. A careful review of relevant literature clearly suggests that drought-induced tree mortality has been widely studied, but not human deaths; this is a fertile area for future research. Not pursued in this book due to space limitations, temporal disaster death trends, preferably for an extended period, should be examined by all disaster together or individually by type at the country level. This type of investigation is necessary because disaster death risk differs from one country to another. For example, analogous to the world in general, earthquakes accounted for the most significant percentage of the death toll among all natural disasters in China, while in the United States, extreme cold or heat accounted for more than two-thirds of the total number of deaths attributed to natural events (see Thacker et al. 2008). Despite the availability of secondary data, subnational level studies of disaster trend analyses are also lacking. The national and subnational level study will provide insights for national and subnational authorities to prevent disaster deaths regarding what has happened in the past and what is currently happening to reduce mortality. In essence, the analysis of past disasters provides essential clues to reducing mortality in future events. Unique patterns of death have been noted among different types of disasters.

Research on the handling of bodies resulting from death by natural disasters is limited, particularly in developing countries. The 2004 IOT marked the beginning of such research in these countries. More research is needed to examine the nature of improvements these countries have made since 2004. At that time, the management of corpses was exceptionally deficient in the affected countries principally because of the utilization of untrained personnel to handle them at all stages (Gupta 2009). Further research is needed to understand how to deal with unidentified bodies following a mass-fatality disaster.

Like disaster death trends, there is a research gap regarding the causes and circumstances of deaths caused by several natural disasters such as lightning, heatwaves, ice and snowstorms, and volcanic eruptions. In general, more studies are needed to analyze the causes and circumstances of disaster deaths by using gender perspective and disaggregating by demographic and socio-economic conditions. Chapter 4 suggests that many deaths by drowning are caused by multiple disasters (all type of floods, tsunamis, and tropical cyclone and associated storm surges). Research should be conducted on how to reduce drowning-related disaster deaths.

While risk factors associated with deaths caused by cyclones, earthquakes, floods, and tornadoes were relatively widely studied globally, little attempt has been made to examine factors related to other natural disasters such as blizzards, droughts, extreme heatwaves, and wildfires. Hazards researchers should seriously consider this research gap. Also, relevant research is lacking in some regions or countries of the world. For example, research conducted to explore the determinants of deaths caused by heatwaves in Sub-Saharan African countries is very sparse. Although studies are available on the determinants of disaster deaths, little attempt has been made to identify which broad factors are more influential in causing disaster deaths, that is, do hazard factors contribute more, or do vulnerability factors weigh more heavily? This is another area of promising research opportunity related to natural disasters.

Most of the suggested future research on disaster deaths should be studied using a gender perspective. Gender-disaggregated mortality data will provide clues to implement actions and investments designed to improve equality and reduce disaster risk for all members of a society. The Sendai Framework of the United Nations also emphasized on gender-sensitive disaster risk reduction policies, plans, and programs. A similar perspective is also applied to disaggregate mortality data by demographic and socio-economic conditions of the disaster victims.

The application of Geographical Information Systems (GIS) techniques, Global Positioning System (GPS), and remote sensing has been used in hazards mapping and assessment, disaster management, and rescue operations and emergency relief distribution (e.g., Curtis and Mills 2009; Boyd 2010). While these advanced and modern techniques, along with web-based technologies, provide a real-time and dynamic way to represent spatial and temporal information on disaster mortality, their use in disaster deaths studies is limited. Several studies (e.g., Feng et al. 2013, 2014) have applied high-resolution remote sensing and geoinformation systems to estimate the casualty rate in selected earthquake events. One of the key reasons for limited studies is the unit of study. Death data are collected and traditionally analyzed at individual level, ignoring spatial analyses of various components of disaster deaths. Disaster mortality, its trends, causes, and determinants, is studied at primarily continent and country levels, but hardly at subnational levels such as census tract and block levels. Note that these technologies can be used in nonspatial context as well.

A better understanding at different subnational level analyses of disaster-induced deaths is crucial for implementing effective and evidence-based policies and programs for reducing such mortality. Spatial analyses can help local disaster managers, planners, and administrators efficiently allocate scarce financial, human, and technical resources to lower deaths from natural disasters (Aksha et al. 2017). Such analyses not only provide insights regarding reducing human, economic, and environmental losses and damage, but also have potential to suggest which mitigation measures are appropriate at what spatial levels to prevent loss of lives from all natural disasters. Applying GIS and remote sensing techniques allows us to explore different mortality-related spatial patterns, to test for statistically significant spatial clusters of fatalities, and to identify hot or low spots, and temporal and multiscalar dynamics.

Notes

1. In developed countries, people often voluntarily evacuate from areas outside a declared evacuation area, causing road congestion and delays in reaching a safe destination; this is called a "shadow evacuation," which should be discouraged (Dow and Cutter 2002).
2. Paul and Dutt (2010) reported that cyclone shelters in some coastal areas did not function as planned because some of them had been used as a place of defecation and others as a cowshed.
3. The shelter's basic component is the construction of a small house of durable cyclone-resistant materials attached to the main dwellings (Diacon 1992).
4. Post-Hurricane Katrina, the United States has established pet shelters where evacuees can keep their pets before hurricanes and other natural disasters. There were other reasons (e.g., taking care of disabled family member, elderly, no car, or no money for costs of transportation to leave the area or motel room) for decision to stay home in New Orleans during Hurricane Katrina. The event occurred at the end of month when people had not enough money.
5. Only around 7 percent of homes in the United States are mobile homes (Simmons and Sutter 2011).

References

Ahsan, M.N., K. Takeuchi, K. Vink, and M. Ohara. 2016. A Systematic Review of the Factors Affecting the Cyclone Evacuation Decision Process in Bangladesh. *Journal of Disaster Research* 11(4): 742–753.

Aksha, S.K., L. Juran, and L.M. Resler. 2017. Spatial and Temporal Analysis of Natural Hazards Mortality in Nepal. *Environmental Hazards*. http://doi.org/10.1080/17477891.2017.1398630.

Alexander, D.A. 1993. *Natural Disasters*. New York, NY: Chapman & Hall.

Boyd, E.C. 2010. Estimating and Mapping the Direct Flood Fatality Rate for Flooding in Greater New Orleans Due To Hurricane Katrina. *Risk, Hazards & Crisis in Public Policy* 1(3): 91–114.

Brown, S.K., S.F. Jenkins, R. Stephen, H. Odbert, and M.R. Auker. 2017. Volcanic Fatalities Database: Analysis of Volcanic Threat with Distance and Victim Classification. *Journal of Applied Volcanology* 6: 15. http://doi.org/10.1186/s13617-0067-4.

CDC (Centers for Disease Control and Prevention). 2017. *Death Scene Investigation After Natural Disasters or Other Weather-Related Events Took Kit: First Edition.* Atlanta, Georgia: CDC.

Ching, P.K., V.C. de los Reyes, M.N. Sucaldito, and E. Tayag. 2015. An Assessment of Disaster-Related Mortality Post-Haiyan in Tacloban City. *Western Pacific Surveillance and Response Journal* 6(31): 34–38.

Cole, T.W., and K.L. Fellows. 2008. Risk Communication Failure: A Case Study of New Orleans and Hurricane Katrina. *Southern Communication Journal* 73(3): 211–228.

Cross, R. 2015. Nepal Earthquake: A Disaster that Shows Quakes does not Kill People, Buildings Do. *The Guardian*, 30 April.

Curtis, A., and J.W. Mills. 2009. *GIS, Human Geography, and Disasters.* San Diego, CA: University Readers.

Diacon, D. 1992. Typhoon Resistant Housing in the Philippines: The Core Shelter Project. *Disasters* 16(3): 266–271.

Doocy, S., A. Daniels, S. Murray, T.D. Kirsch. 2013. The Human Impact of Floods: A Historical Review of Events 1980–2009 and Systematic Literature Review. *PLOS Currents Disasters*, April 16: 5. https://doi.org/10.1371/currents.dis.f4deb457904936b07c09daa98ee8171a.

Dow, K., and S.L. Cutter. 2000. Public Orders and Personal Opinions: Household Strategies for Hurricane Risk Management. *Environmental Hazards* 2: 143–155.

Duclos, P.J., L.M. Sanderson, and K.C. Klontz. 1990. Lightning-Related Mortality and Morbidity in Florida. *Public Health Report* 105(3): 276–282.

Feng, T.Z. Hong, H. Wu, Q. Fu, C. Wang, C. Jiang, and X. Tong. 2013. Estimation of Earthquake Casualties using High-Resolution Remote Sensing: A Case Study of Dujiangyan City in the May 2008 Wenchuan Earthquake. *Natural Hazards* 69: 1577–1595.

Feng, T., Z. Hong, Q. Fu, S. Ma, X. Jie, H. Wu, C. Jiang, and X. Tong. 2014. Application and Prospect of a High-Resolution Remote Sensing and Geo-information System in Estimating Earthquake Casualties. *Natural Hazards and earth Sciences* 14: 2165–2178.

Green, H.K., O. Lysaght, D.D. Saulnier, K. Blanchard, A. Humphery, B. Fakhruddin, and V. Murray. 2019. Challenges with Disaster Mortality Data and Measuring Progress Towards the Implementation of the Sendai Framework. *International Journal of Disaster Risk Science* 10: 449–461.

Gupta, K. 2009. Cross-Cultural Analysis of Response to Mass Fatalities following 2009 Cyclone Aila in Bangladesh and India. *Quick Response Report #216.* Hazards Center, University of Colorado at Boulder.

Haque, C.E., and D. Blair. 1992. Vulnerability to Tropical Cyclones: Evidence from the April 1991 Cyclone in Coastal Bangladesh. *Disasters* 16(3): 217–229.

Haque, U., M. Hashizume, K.N. Kolivras, H.J. Overgaard, B. Das, and T. Yamamoto. 2012. Reduced Deaths Rates from Cyclones in Bangladesh: What More Needs to be Done? *Bulletin of World Health Organization* 90(2): 150–156.

Hettiarachchi, S.S.L., and W.P.S. Dias. 2013. The 2004 Indian Ocean Tsunami: Sri Lankan Experience. In *Natural Disasters and Adaptation to Climate Change*,

edited by Boulter, S., J. Palutikof, D.J. Karoly, and D. Guitart, pp. 158–166. New York: Cambridge University Press.
Ikeda, K. 1995. Gender Differences in Human Loss and Vulnerability in Natural Disasters: A Case Study from Bangladesh. *Indian Journal of Gender Studies* 2(2): 171–193.
Matthies, F., G. Bickler, N.C. Marin, and S. Hales (eds). 2008. *Heat-Health Action-Plan – Guidance*. Copenhagen: WHO Regional Office for Europe.
Mersereau, D. 2013. Why So Many People Died from Haiyan and Past Southeast Asia Typhoon, 11 November. *The Washington Post* (www.washingtonpost.com/news/capital-weather-gang/wp/2013/11/11/inside-the-taggering-death-toll-from-haiyan-and-other-southeast-asia-typhoons/ – last accessed 6 August, 2019).
Nadiruzzaman, M., and B.K. Paul. 2013. Post-Sidr Public Housing Assistance in Bangladesh: A Case Study. *Environmental Hazards* 12(2): 166–179.
Paul, B.K., and D. Che. 2011. Opportunities and Challenges in Rebuilding Tornado-Impacted Greensburg, Kansas as "Stronger, Better, and Greener." *GeoJournal* 76(1): 93–108.
Paul, B.K., and S. Dutt. 2010. Hazard Warnings and Responses to Evacuation Orders: The Case of Bangladesh's Cyclone Sidr. *Geographical Review* 100(3): 336–355.
Paul, B.K., H. Rashid, M.S. Islam, and L.M. Hunt. 2010. Cyclone Evacuation in Bangladesh: Tropical Cyclones Gorky (1991) vs. Sidr (2007). *Environmental Hazards* 9: 89–101.
Paul, B.K., and M.K. Rahman. 2013. Recovery Efforts: The Case of the 2007 Cyclone Sidr in Bangladesh. In *Natural Disasters and Adaptation to Climate Change*, edited by Boulter, S., J. Palutikof, D.J. Karoly, and D. Guitart, pp. 167–173. New York: Cambridge University Press.
Plumer, B. 2014. Tornadoes are Inevitable – So How Do We Stop Them from Killing People? *Vox* (https://www.vox.com/2014/4/29/5664758/what-can-we-do-to-make-tornadoes-less-deadly – last accessed 16 April, 2020).
Schencking, C. 2013. *The Great Kanto Earthquake and the Chimera of National Reconstruction in Japan*. New York: Columbia University Press.
Simmons, K.M., and D. Sutter. 2011. *Economic and Societal Impacts of Tornadoes*. Boston: American Meteorological Society.
Thacker, M.T.F., R. Lee, R.I. Sabogal, and A. Henderson. 2008. Overview of Deaths Associated with Natural Events, United States, 1979–2004. *Disasters* 32(2): 303–315.
van Duin, M.J., and M.J. Bezuyen, M.J. 2000. Flood Evacuation during the Floods of 1995 in the Netherlands. *Floods*, Vol. 1, edited by Parker, D.J., pp. 350–360. London: Routledge.

Index